Modulation, Transmission and Antenna Technology in Wireless Communications

Modulation, Transmission and Antenna Technology in Wireless Communications

Edited by **Timothy Kolaya**

WILLFORD PRESS

New York

Published by Willford Press,
118-35 Queens Blvd., Suite 400,
Forest Hills, NY 11375, USA
www.willfordpress.com

Modulation, Transmission and Antenna Technology in Wireless Communications
Edited by Timothy Kolaya

International Standard Book Number: 978-1-68285-055-8 (Hardback)

Printed in the United States of America.

Contents

Permissions

List of Contributors

Preface

The world is advancing at a fast pace like never before. Therefore, the need is to keep up with the latest developments. This book was an idea that came to fruition when the specialists in the area realized the need to coordinate together and document essential themes in the subject. That's when I was requested to be the editor. Editing this book has been an honour as it brings together diverse authors researching on different streams of the field. The book collates essential materials contributed by veterans in the area which can be utilized by students and researchers alike.

Wireless technology has taken the world by storm in the recent past. It is a rapidly developing technology which has applications in our day to day lives. This book aims to delve deeper into various researches and technological advancements in wireless communications. It discusses significant concepts such as types of wireless communications, signal processing algorithms, signal modulation, etc. Coherent flow of topics and extensive use of examples make this book an invaluable source of knowledge. This book is highly recommended for students pursuing electronics and communication and allied fields. New researchers will also find this book very useful as it will help them by foregrounding their knowledge in this branch.

Each chapter is a sole-standing publication that reflects each author's interpretation. Thus, the book displays a multi-facetted picture of our current understanding of applications and diverse aspects of the field. I would like to thank the contributors of this book and my family for their endless support.

Editor

Implementation of Channel Estimation and Modulation Technique for MIMO System

Mrs. VEENA M.B.[1] & Dr.M.N.SHANMUKHA SWAMY[2]

Research scholar [1] , Professor[2]
ECE department, SJCE, Mysore.
Karnataka, India
e_mail: veenahod@rediffmail.com [1] , mnsjce@yahoo.co.in[2]

Abstract

Future wireless communication system have to be designed to integrate features such as high data rates, high quality of service and multimedia in the existing communication framework. Increased demand in wireless communication system has led to demand for higher network capacity and performance. Higher bandwidth, optimized modulation offer practically limited potential to increase the spectral efficiency. Hence MIMO systems utilizes space multiplex by using array of antenna's for enhancing the efficiency at particular utilized bandwidth. MIMO use multiple inputs multiple outputs from single channel. These systems defined by spectral diversity and spatial multiplexing. The aim of this paper is to design and implement of channel estimation method and modulation technique for MIMO system. The design specifications are obtained using MATLAB. The RTL coding is carried for the design to be implemented on Xilinx FPGA.

Key words: MIMO, FPGA, Space multiplex, RTL.

1. Introduction

MIMO describes the ways to send data from multiple users on the same frequency/time channel using multiple antennas at the transmitter and receiver end. A transmitter/receiver system uses multiple antennas not only for transmitting data between corresponding antennas but also between adjacent antennas. The data is received in the form of MIMO Channel Matrix. A top level MIMO system is shown in Fig.1 MIMO system is used in many applications like WiMax, WiFi, WLANs, and many more signal processing applications.

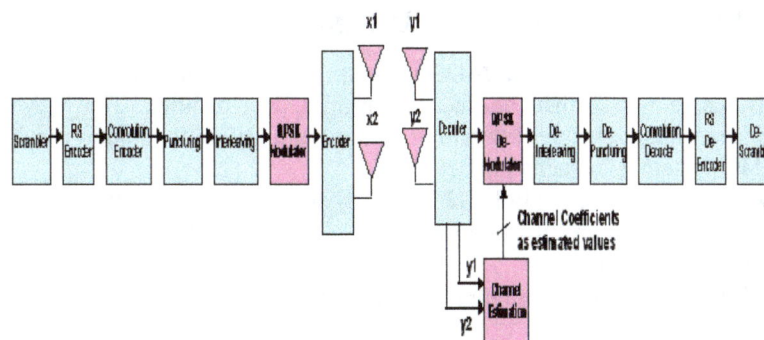

Figure 1 : MIMO System [1]

Blocks in MIMO system are scrambler/descrambler, RS encoder/decoder, convolution encoder/decoder, puncturing/de-puncturing, interleaver/de-interleaver, modulator/ demodulator, diversity or space time encoder/decoder and channel estimation. This paper aims at design and implementation of a simple and most efficient channel estimation method and a good modulation technique for increasing the channel capacity, bandwidth, increasing bit rates and eliminates intersymbol interference.There are well-known training based channel estimation methods are; least square (LS), Minimum Mean Square Estimation (MMSE), Least mean square (LMS), Recursive Least Square (RLS). In this paper the main focus is on simple LS algorithm, which is simple to analyse and efficient. The comparison of LS is not performed. The main aim is to reduce the computational complexity of channel estimation using LS algorithm and implementing 2x2 MIMO system using QPSK modulation technique.

2. Literature Survey

Wout Joseph, et.al. [1] proposed an algorithm for the performance study on IEEE 802.16-2004 for different conditions like for different channel models and MIMO system. Chia-Liang Liu [4] explained the impact of I/Q imbalance on QPSK-OFDM-QAM Detection. The in-phase and the quadrature phase components are very important component in QPSK. Yantao Qiao, et al.[5], has made a research on an iterative algorithm of least square channel estimation in MIMO OFDM systems. The main objective of this paper is an iterative channel estimation algorithm for MIMO OFDM is proposed. Sarod Yatawatta, et.al.[7] proposed a solution for minimizing the energy spent on during the channel estimation when subjected to known error and delay when timing symbols are transmitted. The minimization of energy is carried by reducing the hardware, also by using a low rank equalization at the receiver. Benoit Le Saux, et.al, [8] proposed a MIMO system with OFDM has greater potential like reduction in inter-symbol interference, decrease in fading, increase in bandwidth, and increase in data rates. The performance of MIMO system degrades due to inaccurate channel estimation over frequency selective fast-varying channels.

Riza Abdolee, et.al.[10] proposed a method to reduce the computational complexity of channel estimation algorithm for MIMO-OFDM. Channel estimation is high intensive which suffer from high computational complexity. Solution for high efficient channel estimation and simplified computational complexity is stated. Deseada Bellido, et.al, [11] proposed LS channel estimation algorithm for MIMO-OFDM. This evaluation has been made using pilot design rules that guarantee a bounded error level for the estimation. This method is used for estimation of the channel matrix. Markus Myllyla, et.al, [12] proposed a method for performance evaluation of 2 FPGA implementation of a LMMSE based detector for radio channels.

3. Channel Estimation and QPSK Design

In this 2x2 MIMO system is designed. MIMO system [3] has multiple transmitter antennas and multiple receiver antennas so that the data is transmitted in parallel.. The MIMO system is designed with least square channel estimation method and QPSK modulation technique. Each module in MIMO system is explained in detail.

2x2 MIMO system is shown below in Fig.2 MIMO system has following blocks.

1) Least Square (LS) channel estimation block
2) QPSK modulator/demodulator
3) 2x2 Ideal MIMO Channel (2 transmitter & 2 Receiver antennas).

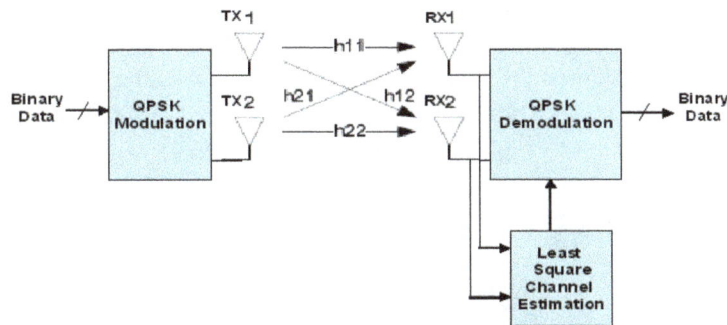

Figure 2 : 2x2 MIMO system

3.1 Least Square Channel Estimation Method

In communication systems, channels are usually multi-path channels, which cause inter-symbol interference in the received signal. Channel estimators[5] require the channel impulse response (CIR). The channel estimation is based on the known sequence of bits called training sequence which is unique for each transmitter. Here the known training sequence is transmitted so that the channel coefficients are obtained. There are different standards used for transmitting training sequence like IEEE 802.16 standard.

3.1.1 Channel estimator for single transmitter and single receiver

In any communication system noise get added in the channel with the signal transmitted. The digital signal transmitted over a faded multi-path channel *'h'*. The noise get added which is modeled as additive white Gaussian noise *'n'* as shown in Fig.3 The demodulation problem is to detect the transmitted bits *'x'* from the received signal *'y'*. The received signal also has channel coefficients multiplied. The detector needs these channel estimates for that specific channel and channel estimation device. The system is shown by the equation (3.1).

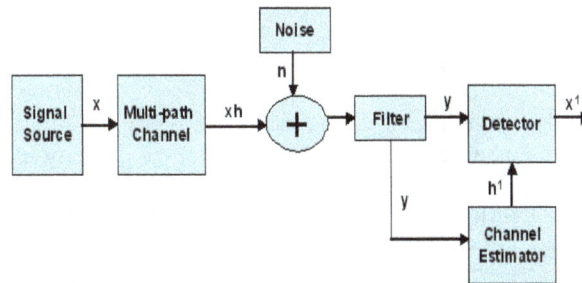

Figure 3 : Block diagram of a noise corrupted system

The received signal y is given as

$$y = Mh + n \qquad (3.1)$$

$$h = [h_0 \, h_1 \, ...h_L]^T \qquad (3.2)$$

In each transmission burst the transmitters sends a unique training sequence which has elements $m_i \in \{-1,+1\}$. Here M is the matrix for training sequence. Hence the final least square channel estimation equation is given by equation (3.3).

$$h^1_{LS} = (M^H M)^{-1} M^H y \qquad (3.3)$$

3.1.2 Channel estimator for 2x2 channels

2x2 joint channel estimation is considered for the MIMO design. The channel is considered to be ideal wherein noise is not taken in account. The sequence *'x'* is taken from 2 different transmitters and 4 channels are considered as shown in Fig.4 From each transmitter unique training sequences are transmitted in concatenation with the QPSK modulated data. The training sequence is transmitted for the identification of the transmitter from which the data is obtained at the receiver end. Here diversity technique is used but the alamouti encoder and decoder is not used.

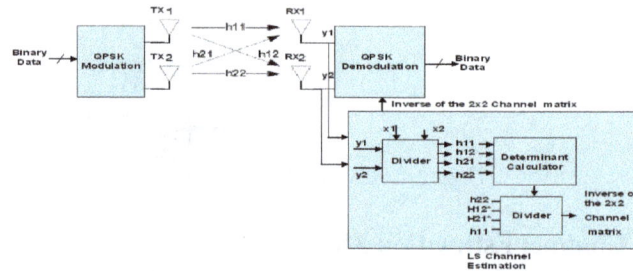

Figure 4 : Block showing QPSK modulator/demodulator with LS channel estimation

The 2x2 channel estimation block diagram is shown in Fig.2. Initially the training sequence is transmitted so that the channel coefficients are calculated. First the transmitter is made active and the training sequence is sent. The tx2 is made inactive i.e. nothing is sent from tx2. The h11 and h12 channel coefficients are obtained. Then the tx1 is made inactive and tr2 is sent through tx2. Then the h21 and h22 coefficients obtained by simplifying the equations (3.3). The data obtained at the receivers rx1 and rx2 are shown by the equations (3.4.and 3.5).

$$y1 = x1 * h11 + x2 * h21 \qquad (3.4)$$
$$y2 = x1 * h12 + x2 * h22 \qquad (3.5)$$

After which the transmitted bits are decoded by multiplying y with the inverse of the 2x2 channel matrix h.

$$h^1 = y/x \qquad (3.6)$$

3.2. QPSK modulator/ demodulator

PSK is a digital modulation technique which is most commonly used modulation technique in present digital communication systems. In PSK modulation, the phase of the carrier is altered in accordance with the input binary coded information. The PSK is further subdivided into BPSK, 8-PSK, 16-PSK, QPSK, DPSK. In binary phase shifting keying the transmitted signal is sinusoid of fixed amplitude, has fixed phase.

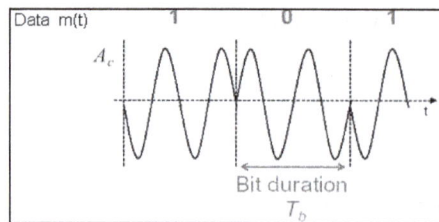

Figure 5 : Phase Shift Keying [1]

QPSK [4] is a phase modulation scheme, used in constellation mapping. Here the input bits stream is converted into complex stream using equation (3.7) and where the I and Q both are in phase with I-out and Q-out respectively. QPSK modulator accepts the binary bits as inputs taken as a symbol and converts them into complex value. QPSK takes only 4 symbols and generate its complex value in this fashion.

$$D = (I +jQ)*K_{MOD} \qquad (3.7)$$

Where $K_{MOD} = 1/1.414$

Table 1 : Table showing the Inputs and Outputs of QPSK Modulator

Input Bits	I-out	Q-out
00	-1	-1
01	-1	+1
10	+1	-1
11	+1	+1

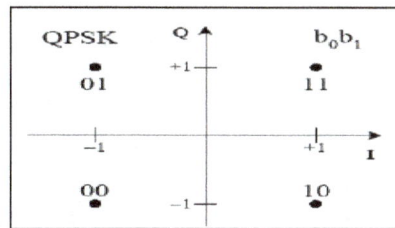

Figure 6 : QPSK Constellation Mapping

3.4. Working of 2x2 MIMO System

The 2x2 MIMO system has different blocks like QPSK modulator, ideal channel, channel estimation block and QPSK demodulation. QPSK modulator takes the binary bits as input. Here the inputs bits are taken as symbols. For example 00, 01, 10,11are 4 inputs symbols for QPSK. The symbols are converted into complex values as shown in table 1. The complex output of the modulator is appended with the training sequence at 2 transmitters. Here the diversity technique is used to send the same data (QPSK modulated output) with different training sequences. The channel coefficients are multiplied with the modulated data. Here 4 paths are considered or 2x2 channels as shown in Fig.7

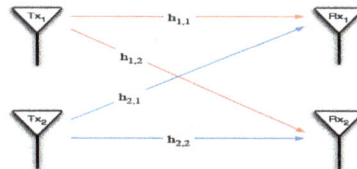

Figure 7 : Communication Channel for 2x2 MIMO System [13]

The channel is an ideal channel i.e. the noise is not considered. The reliability is increased by employing the diversity technique i.e. transmitting the same information across multiple channels. If one of the channels is not used or if the data is lost in space then the information/data can still be recovered from redundant transmission over the channels and hence the reliability of the communication system is improved.

The data transmitted from transmitter tx1 across the channels hand received by both the receiver's rx1 and rx2. Here the use of the available channels is done to increase the capacity and reliability. Then the received sequence is taken into the channel estimation block to find the estimated channel coefficients. The received data from LS channel estimator [5] block are the given to the decision block in the QPSK demodulator block to get back the sent binary sequence. The transmitted sequence is the checked with the received sequence at the QPSK modulator and the Demodulator block respectively.

4. Simulation & Implementation Overview

The MATLAB specifications of MIMO system from MATLAB are shown below.
1) 2x2 MIMO Technique: 2 transmitters and 2 Receiver antennas
2) QPSK modulation technique
3) Number of bits transmitted 64bits
4) Number of iterations = 100
5) Signal to noise ratio SNR = [0:3:9] i.e. 0, 3, 6, 9 in dB

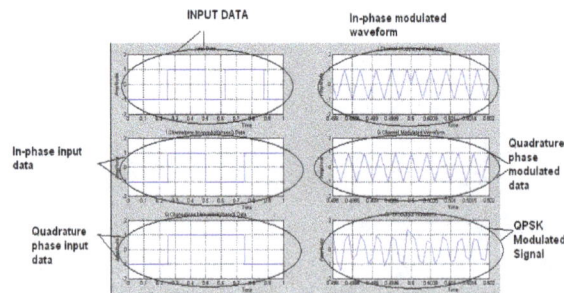

Figure 8 : QPSK modulator MATLAB outputs

QPSK modulated output is shown in Fig.8 Input binary data is sent through QPSK modulator, the data taken into is in the form of symbols. The in-phase and quadrature waveform and the QPSK modulated data are shown in Fig.8 The modulated data sent through the noisy channel as shown in Fig.4 the data get mixed with the channel coefficients and the noise. The bit error rate with respect to signal to noise

ratio the error rate goes on reducing as shown in Fig.9 (a). The error rate goes on reducing from 10^{0} to 10^{-1} as the SNR is from 0 to 10dB. As the number of iterations increases the error at the output which is given as the difference between the input and the output signal becomes as zero. The error plot is shown in Fig.9 (b) is for 100 iterations. There is a mismatch at 2 points and the difference is plotted. These plots as shown in Fig.9 are for LS channel estimation method.

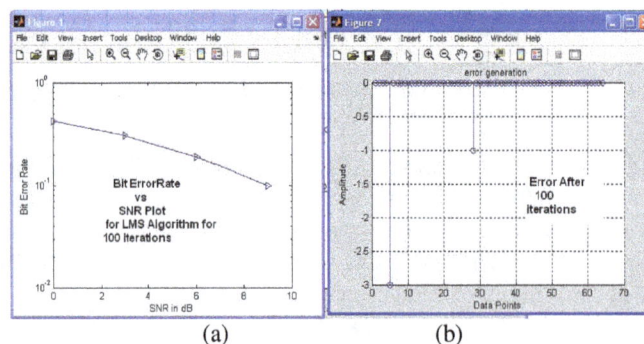

(a)   (b)

Figure 9 : Graph showing BER vs SNR plot and Error generation after 100 iterations

As the number of iterations increase the error reduces so that the BER vs SNR ratio goes on reducing and becomes negligible. Whenever there is a mismatch between the input and the output sequence of MIMO system then the difference is shown as the error.
The plots shown in Fig.10 are the input to a system is a random data generated using 'randn' in MATLAB. The random data generated is fed to the QPSK modulated. The data get modulated and then the modulated output is sent through the QPSK demodulator. The demodulated output

same as input as shown in Fig.10 The plots as shown in Fig.10 are from MATLAB. Here the input is taken as 64 symbols.

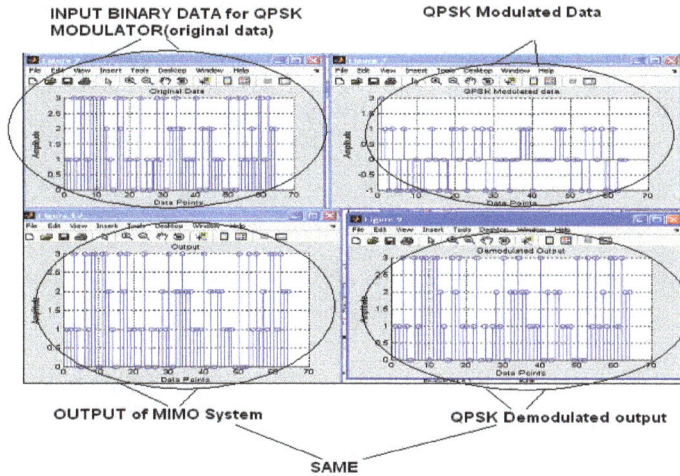

Figure 10 : Plots showing the input modulated output and demodulated output

2x2 MIMO system is designed using Verilog-HDL and simulated in ModelSim. Each block is separately verified for its functionality. Each block is synthesized and the integrated output is also verified by implementing Xilinx FPGA. The top block of 2x2 MIMO system block diagram from Xilinx is shown in Fig.11 All the blocks in 2x2 MIMO system like QPSK modulator/demodulator, ideal channel and the LS channel estimation blocks are integrated and the connections are shown in Fig.11 with inputs and the outputs.

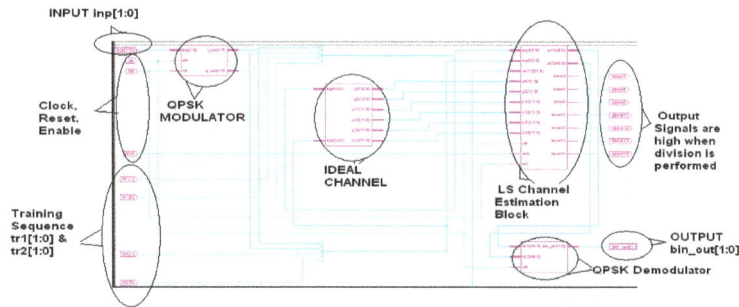

Figure 11 : 2×2 MIMO System in Xilinx

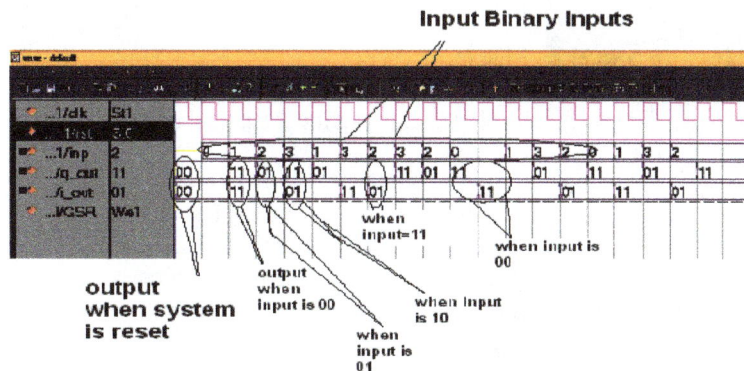

Figure 12: Waveform showing QPSK modulated output from ModelSim

The waveform shown in Fig.12 shows the QPSK modulated output. As explained before each symbol is displayed as complex value. Similarly symbol 00 gives in phase value as 11 and quadrature phase value as 11 as shown in Table 2.

Table 2: Table showing Binary Symbol representation as i_out and q_out

Binary Symbol	I_out	Q_out
00	11	11
01	11	01
10	01	11
11	01	**01**

The Fig.13 shows the waveform for ideal channel. This block is a completely combinational block. Here the data from QPSK is passed so that the channel coefficients are multiplied with it. Here as explained before the training sequence transmitted from tr1 is 01{+1} and from transmitter 2 is 11 {-1}.

Table 3: Table Displaying the Demodulated Output

x11	X12	Output of QPSK demodulator
65	67	00 => 0
56	58	01 => 1
31	33	10 => 2
22	24	11 => 3

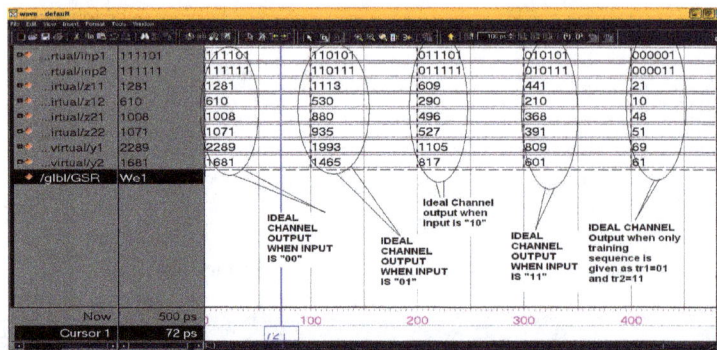

Figure 13: Waveform showing ideal channel outputs

The channel coefficients are taken from the MATLAB and are multiplied by 2^{10} i.e. the fixed point conversion is done for the channel coefficients. The inp1 is multiplied with the h11, h12 and inp2 is multiplied with the h21 and h22. Then the z11 is added with z21 and z12 added with z22 is to get y1 and y2 respectively. The y1 and y2 are received at the receiver. For different input sequences the waveform is shown in Fig.13

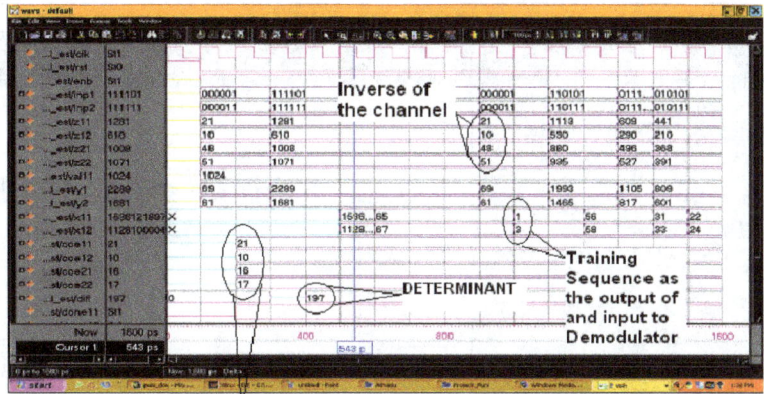

Figure 14: Waveform showing the LS channel Estimation block outputs

The outputs of the ideal channel are fed as inputs to the LS channel estimation where the inverse of channel coefficients is calculated. The channel is 2x2 and the matrix is 2x2 square matrix. Then the inverse of the channel matrix is multiplied with the received sequence. As shown in equation (2.6). The Fig.14 shows the LS channel estimation outputs wherein the inverse of the square matrix is calculated. The determinant is calculated for inverse matrix calculation.

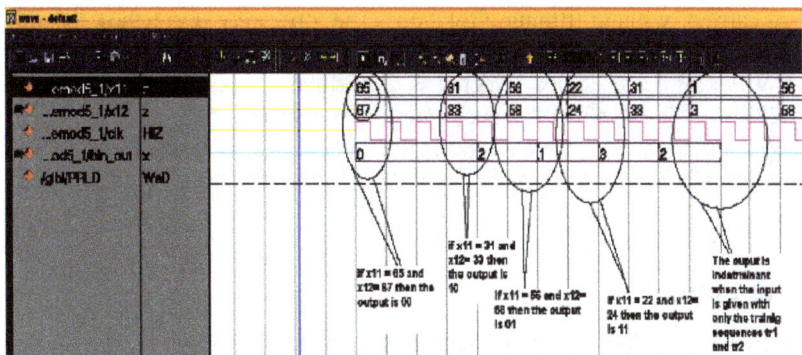

Figure 15 : Waveform showing QPSK demodulated output

The outputs of the LS channel estimation block are fed to the decision device used in QPSK demodulator which defines the value x11 and x12 as the values. The values in the required range i.e. the value x11 and x12 are 31 and 33 then the output should be '10' i.e. 2 as shown in Fig.15 Similarly, for different values of x11 and x12 the outputs of QPSK demodulator are shown in Table 3.

The integrated 2x2 MIMO system design output is as shown in Fig.16 the waveform shows the 2x2 MIMO system output. Here the clock is applied and the reset is made high then the system is made reset. Then the reset signal is made low and the enable signal is made high. The inp[1:0] is the given the value as the required symbol to transmit. The training sequence tr1 is given as +1 and training sequence tr2 is given as -1. Then the output bin_out[1:0] is obtained as the input inp[1:0]

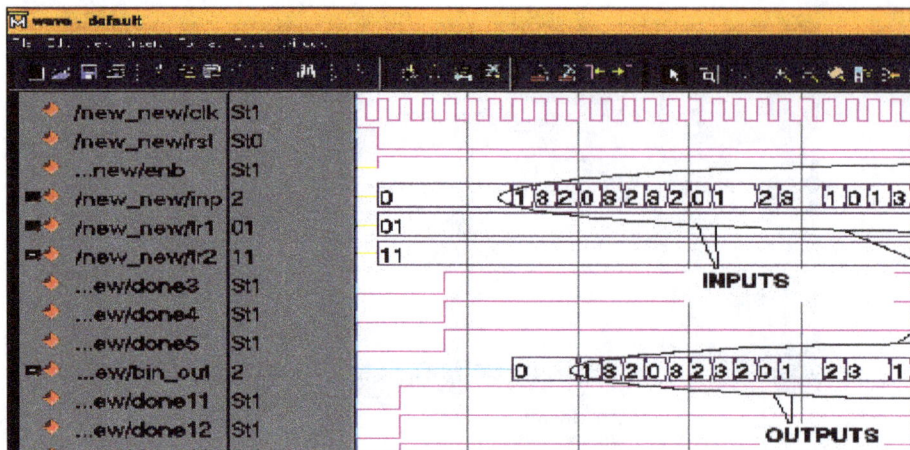

Figure 16.   2x2 MIMO System output waveform in ModelSim

5. Conclusion

The MIMO system design is simulated in MATLAB to arrive at the specifications. The RTL code is successfully simulated in ModelSim. The design synthesized and implemented on Virtex2Pro FPGA board and the results were validated using Logic Analyzer. The synthesis and timing is verified and the timing is met for both setup and hold in DC and PT. DFT is also carried without timing violation. The operating frequency of the design is 13MHz obtained from Xilinx. The top design takes about 3999 number of slices out of 4928 slices i.e. in Virtex2Pro the device selected is 2vp7ff896 at speed grade of 6 with operating frequency 7.27MHz. The 2vp30ff896 is selected with a speed grade of 6, the slices are 1902 out of 13696 i.e. 13% resources usage and operating frequency is 13.388MHz and minimum period of 74.691ns. Timing verified is all met with positive slack with zero violations.

Acknowledgement

Authors would like to thank the Management Vemana Institute of Technology, KRJS, Bangalore, for funding and providing necessary help to complete this paper.

References

[1] Joseph, W. Reynders, W. Debruyne, J. and   Martens, L., "*Influence of Channel Models and MIMO on the Performance of a System based on IEEE 802.16*",Wireless Communications and Networking Conference, ISBN 1-4244-0659-5, pp.1826-1830, 11-15 March, 2007.

[2] Simon S Haykin and Michael Moher, "*Modern Wireless Communication*", Second Edition, Prentice Hall publication 2004, ISBN-13: 9780130224729.

[3] George Tsoulos, "*MIMO System Technology For Wireless Communications*", Revised Edition, CRC Publisher, 2006.

[4] Chia-Liang Lui, "*Impacts of I/Q Imbalance on QPSK –OFDM-QAM Detection*", IEEE Transactions on Consumer Electronics, vol. 44, no. 3, pp. (984-989), August, 1998.

[5] Yantao Qiao, Songyu Yu, Pengcheng Su and Lijun Zhang, "*Research on an Iterative Algorithm of LS Channel Estimation in MIMO OFDM Systems*", IEEE Transactions on Broadcasting, vol. 51, no. 1, pp. (149-153), March 2005.

[6] S.Cui, A.J.Goldsmith and A.Bahai, "*Energy Efficiency in MIMO and cooperative MIMO techniques in Sensor Networks*", IEEE Journal on Selective Areas in Communication, vol. 22, no. 6, pp. (1089-1098), August 2004.

[7] Yatawatta,S, Petropulu,A.P and Graff,C.J, *"Energy efficient Channel Estimation in MIMO Systems"*,IEEE international conference on Acoustics,Speech and Signal processing, vol.4, no.1, pp.(317-320), 18-23 March,2005.

[8] Saux,B.L and Helard,M., *"Iterative Channel Estimation based on Linear Regression for a MIMO-OFDM System"*,IEEE interanational Conf. on Networking and Communications,vol.1, no.1, pp.(356-361),19-21June, 2006.

[9] Changchuan Yin, Jingyu Li, Xiaolin Hou and Guangxin Yue, *"Pilot Aided LS Channel Estimation in MIMO-OFDM Systems"* The 8th International Conference on Signal Processing,vol.3, pp.(16-20), 2006.

[10] Reza Abdotee, Tharek Abd.Rehman and Savia Mahdaliza Idrus, *"Computational Complexity Reduction for MIMO-OFDM Channel Estimation Algorithm "*, elecktika Journal of electrical engineering , vol.9,No.1, pp.(30-36) , 2007.

[11] Bellido, Deseada, Entrambasaguas and Jose T. *"MSE evaluation at reception end in MIMO-OFDM systems using LS channel estimation"*, Waveform Diversity and Design Conference, vol.1, pp. (174-177), 4-8 June,2007.

[12] Markus Myllyla, Markku Juntti, Matti Limingoja,Aaron Byman and Joseph,R. Cavallaro *"Performance Evaluation of Two LMMSE Detectors in a MIMO-OFDM Hardware Testbed"*, Conf.on Signals, systems and Computers, Fortieth Asilomar Pacific Grove, pp. (1161-1165), Oct29-Nov1 2006.

[13] Ian Griffiths *"FPGA Implementation Of MIMO Wireless Communications System"*, University Of New Castle, Australia, 1st November, 2005.

[14] Bernard Sklar *"Digital Communications–Fundamentals and Application"* Published by PEARSON Education. Year 2003.

[15] G.Proakis, Masoud Salehi *"Communication System Engineering"*, Second edition by John. Published by Pearson Education. 2002.

[16] John G. Proakis, Dimitris G. Manolakis, *"Digital Signal Processing Principles Algorithms and Applications"*, 4th Edition, Boston: McGraw Hill.

[17] Himanshu Bhatnagar – *"Advanced ASIC Chip Synthesis Using Synopsys® Design Compiler™ Physical Compiler™ and PrimeTime®"*, Kluwer academic publishers, Second edition – 2002.

[18] David Garrett, Linda Davis, Stephen ten Brink, Bertrand Hochwald and Geoff Knagge, *"Silicon Complexity for Maximum Likelihood MIMO Detection Using Spherical Decoding"*,IEEE journal of Solid State Circuits, vol.39, no.9, pp. (1544-1552), September, 2004.

[19] Yvo L.C. de Jong and Tricia J. Willink , *"Iterative Tree Search Detection for MIMO Wireless Systems"*, IEEE Transactions on Communications, vol. 53, no.6, pp. (930-935), June, 2005.

Veena M.B.: received the B.E.,degree in Electronics & communication from Mysore university in 1991 and the M.E.,degree in Electronics & communication from Bangalore university in 2003 .Currently pursuing Ph.D from V.T.U., Belgaum, India.
Presently working as a Assistant professor, Dept. of Telecommunication Engineering, Vemana Institute of Technology,Bangalore,India. & research interest in the field of MIMO wireless communication system, circuits & systems, Signal processing & VLSI.

M.N.Shanmukha swamy : received the B.E. & M.E.,degree in Electronics & communication from Mysore university. Ph.D from IISC, Bangalore, India. Presently working as a professor, Dept of Electronics & communication Engineering, SJCE, Mysore, India. & research interest in the field of wireless communication system, circuits & systems, VLSI. Adhoc networks & Image processing. Under his guidance many research students are working.

Extended Ad Hoc on Demand Distance Vector Local Repair Trial for MANET

P. Priya Naidu[1] and Meenu Chawla[2]

[1]Associate System Engineer, IBM Pvt. Limited, India
Priya.cse020@gmail.com

[2]Associate Professor, Department of computer Science MANIT, Bhopal, India
chawlam@manit.ac.in

ABSTRACT

Adhoc On-demand Distance Vector (AODV) is a routing schema for delivering messages in a connected Mobile Adhoc Network (MANET). In MANETs, a set of nodes are used to route the data from source to destination and it is assumed that nodes are distributed over the entire region. Connectivity between any sources to destination pair in the network exists when they are in radio range of each other. The technique used to deal with the issues called Local Repair. Local Repair is an important issue in routing protocol which is needed for minimizing flooding and performance improvement. Local Repair is one of the major issues in the protocol; routes can be locally repaired by the node that detects the link break along the end to end path. Local Repair will increase the routing protocol performance. In this paper, the existing Local Repair Trial method in AODV is extended to achieve broadcasting and minimizing the flooding. The Enhanced protocol first creates the group of mobile nodes then broadcasting can be done and if the link breaks then local repair technique can be applied. In the network the numbers of intermediate nodes are increased by using Diameter Perimeter Model. Enhanced AODV-Local Repair Trial (EAODVLRT) protocol is implemented on NS2 network simulator. Simulations are performed to analyze and compare the behavior of proposed protocol (EAODVLRT) for varying parameters such as size of network, node load etc. Proposed protocol has been compared with the existing AODV-LRT in terms of routing load, Data delivery ratio.

KEYWORDS

AODVLRT, MANET, Local Repair ADHOC network & Perimeter routing.

1. INTRODUCTION

A wireless network is a rising technology that will allow users to access services and information electronically, irrespective of their geographic position. Wireless communication network is a collection of independent devices connected to each other. Some of the advantages of wireless network are it is easily deployable and flexible in nature as compared to wired networks. There are two approaches for wireless communication that is centralized cellular network and decentralized approach (adhoc network). In centralized cellular network each mobile unit is connected to one or more fixed base stations, so that a communication between two mobile stations requires involving one or more base stations. In decentralized approach there may be situations where radio network units or nodes, move in terrain where line-of-sight communication rarely is possible between all nodes and where pre-deployed infrastructure cannot be guaranteed. A mobile ad hoc network (MANET) (Fig. 1) each device (nodes) is dynamically self organized in network without using any pre-existing infrastructure.

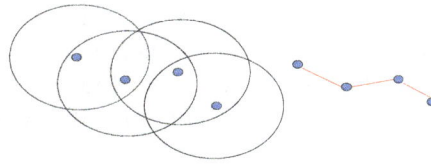

Fig.1 Mobile Ad-Hoc Network

The primary challenge in building a MANET is equipping each device to continuously maintain the information required to properly route traffic. Such networks may be operated by themselves or may be connected to the larger Internet. MANETs are a kind of wireless ad hoc networks that usually have a routable networking environment on top of a link layer ad hoc network. The mobility of nodes in Mobile Ad Hoc networks (MANETs) causes frequent changes of network topologies making routing in MANETs a challenging task.

Motivation for current work is that Ad hoc network allows all wireless devices within range of each other without involving any central access point and administration. Increase in number of nodes degrades the performance of large ad hoc network that makes the design of routing protocols more challenging. There are many simulation study has been done so far for the routing protocols. This paper has been organized as follows: In the following section we briefly review the two protocols AODVLRT (AODV with Local Repair Trials) [1] and AODV [2].

The performance metrics are described based on the comparison of the protocols. Next to this a simulation model has been explained on which basis results are obtained and graphs are generated to compare and analyze the results with the help of performance metrics. We have presented the simulation based comparative performance analysis of routing protocol and finally concluded that the enhanced version of AODVLRT protocol is better under certain traffic conditions and scenarios. The main motivation behind the current work is enhancing the AODVLR protocol by reducing the routing overhead. There will be impact on performance for low bandwidth in wireless link if high routing load is there.

The main objective of this paper is to enhance AODVLRT (AODV Local Repair Trial) protocol by minimizing flooding using perimeter routing. In the previous implementation [1] throughput increases with the increase of routing overhead but, in this paper a novel method is proposed to reduce the two parameters i.e. controlling overhead and increasing throughput are the major areas of focus. The remainder of this paper is organized as follows. In section 2, a short introduction to AODV, AODVLR (AODV Local Repair), AODVLRT (AODV Local Repair Trial) is presented. In Section 3, we suggest an improvement of AODVLRT by implementing perimeter routing. Section 4 describes the simulation model adopted, and then a detailed simulation is performed to evaluate the performance of the Enhanced AODVLRT (EAODVLRT). Conclusions are presented in section 5.

2. AODV AND ROUTE REPAIR

2.1 Overview

Ad hoc on demand distance vector routing is an on demand approach of route finding. Routing can be done when source nodes sends the packet for transmission. AODV differs from the other on-demand routing protocols is in a way that it uses a destination sequence number to determine an up-to-date path to the destination but it doesn't broadcast update information in the network. But in this case the entire topology had being change in the network periodically [3]. Like the on-demand routing AODV source node also floods the route request packet in the network. The routing can be done at the intermediate node is established by comparing the sequence number of source and destination in the route request packets. If a route request is received multiple times, duplicate copies must be discarded. AODV protocol generally uses mobile sensor nodes in a multi- hop wireless network which is quickly adaptable for dynamic link condition, it uses unicast route (asymmetric in nature) to the destination. In AODV the destination node chooses one among all possible discovered routes [3].

Major advantage of AODV is that the connection setup delay is much less than other protocol [3]. While the drawback is that the inconsistent routes are also discovered. The periodic beaconing also leads to unnecessary bandwidth consumption.

2.2 Local Repair AODV

AODV is a popular on demand routing protocols for mobile adhoc network. The major drawback which the AODV suffer lots of link failure [4] with the failure of single node the whole route is rejected AODV can basically work in two repair techniques:

(1) Source Repair   (2) Local Repair

The wireless multi-hop networks are suffered from link failures so, it is necessary to repair the routes. In the Ad-hoc On Demand Distance Vector (AODV) protocol, routes can either be repaired by re-establishing a new route from starting to the source node (Source Repair), or they can be locally repaired by the node that detects the link break along the end-to-end path (Local Repair) [5].AODV is one of the major reactive protocols which mean route discovery are initiated on demand. Once a route is discovered between two nodes, data transfer occurs until the route is broke due to node movement or inference due the erroneous nature of wireless medium. When a route failure happens between two nodes route maintenance can initiated. The upstream node of failure tries to repair that route and this process called local repair [6]. In the AODV routing protocol is reactive protocol which means route discovery can be done on demand, if the particular node failure can occur then the whole routing can be done. To avoid this problem local repair technique is being added by AODV and new protocol had being generated called as AODVLRT [1].

2.3 AODV Local Repair Trial (AODVLRT)

AODVLRT is modification of local repair algorithm used in the route maintenance of the AODV routing protocol. The AODVLRT mainly reduces the routing message overhead resulted from the original AODV local repair algorithm [1]. The enhancement leads to higher throughput and lower latency when compared to AODV. Major difference between AODVLR and AODVLRT is just one trial to find a repair to the route by broadcasting RREQ packet with TTL come from below equation which is taken from [1].

$$TTL = Max\ (0.5 \times N_H,\ TTL_{MNR}) + TTL_{LA}$$

Where,

TTL_{MNR}: the last known hop count from the upstream node of the failure to the destination.
TTL_{LA}: constant value
N_H: the number of hops from the upstream node of the failure to the source of the currently undeliverable packet.

3. IMPROVEMENT TO THE STANDARD AODVLRT

Routing can be done from source node to the destination node by flooding the route request packet. It employs destination sequence numbers to identify the most recent path. The destination sequence number is created by the destination that is included along with any route information it sends to the requesting nodes. Destination sequence number gives the choice between two routes; a requesting node is required to select the one with the greatest sequence number. During the process of routing failure, of a node causes the whole route to be rejected. To overcome this, repairing technique can be used. The behavior of AODV in case of link failure as defined in [1, 4]. In EAODVLRT flooding can be minimized by combining the concept of perimeter routing [7].

3.1 System Model

We represent a wireless ad hoc network by a graph G= (V, E) where V is the set of vertices which represents mobile nodes and E subset of V^2 the set of edges between these vertices. An edge exists between two nodes if they are able to communicate to each other, that is two nodes u and v can communicate if they are in communicating radius of each other. If all nodes have the same range R, the set E is defined as:

$$E = \{(u, v) \in V^2 \mid u \neq v \ \& \ d(u, v) \leq R$$

D (u, v) being the Euclidean distance between u and v. we also define neighborhood set N (u) of the vertex u as

$$N(u) = \{v \mid (u, v) \in E\}$$

This system environment also makes following assumptions:

- Nodes are being connected with the symmetric link properties.
- The set of all nodes in the system is denoted as M= {M1, M2... MN}, where N is the total number of nodes in the network.
- Nodes are dynamic in nature, so that TTL_Threshold value should be initialized dynamically.
- Intermediated nodes between sources to destination nodes can be 40 for using network diameter model.
- Time to live increment (TTL_increment) which is assumed two and TTL_LA values are pre initialized are static. Every node has different battery power.
- Mobility random mobile model is used with predefined pause time.

3.2 EAODVLRT Algorithm

- In the algorithm, an AODV-node informs its neighbours about its own existence by constantly sending ``hello messages" at a defined interval.
- Discovery of neighboring nodes is done by perimeter routing protocol. Through perimeter routing, the sender can only broadcast the RREQ packets to the outer boundary in counter clockwise direction. A RREQ contains the sender's address, the address of the source node and the maximum sequence number received from the node which exists.
- If the source cannot find the destination then that route can be discarded, again new route can be searched by using the local repair techniques.

- Local Repair will increase the routing protocol performance. The major idea is controlling messages from neighbour nodes; this can be done by minimizing flooding.
- In the AODV model, the inquiry about the particular route from source to destination by default is 2 but it can be increased to 7 times in EAODVLRT.
- The receiving node checks whether it has a route to the particular node. If a route exists and the sequence-number for new received route is higher than the existing route then it accepts the new route. The node replies to the requesting node by sending a route reply (RREP). On the other hand, if a route does not exist then the receiving node sends a RREQ by itself in order to find a route for the requesting node.
- If the original node does not receive an answer within a time-limit the node can assume that the source nodes are unreachable. Then the request was sent to all neighbouring nodes which are easily separated by the sequence numbers. Nodes along the route will keep their routing table updated. Otherwise, the nodes will discard the entries after a particular time. To be sure that the route still exists, the sender has to keep the route alive by periodically sending packets. The broken link will send the error message (RERR) to the closest node, so that the nodes which identified the broken link can start to search for a new route.

3.3 Description EAODVLRT Algorithm

In a wireless network, a route is searched from source to destination by broadcasting the route request message by the source node. The broadcasting can be done by using the perimeter routing [7, 8]. Through the perimeter routing the sender can only broadcast the RREQ packet to the outer boundary in counterclockwise direction. When broadcasting is done in perimeter mode, overlapping of links between the neighbor nodes can be avoided by constructing a planer graph using RNG or GG [9, 10]. By using this procedure flooding can be minimized. After flooding, destination node replies with the corresponding route reply message. During this routing process intermediate nodes create routing table entries for both source and destination nodes. There by creating bidirectional end-to-end route. The major drawback of AODV is that it suffers with a lot of link failures [11]. To overcome this problem local repair technique can be used.

In local repair technique once a route is discovered between two nodes, data transfer occurs until the route is broken due to node movement or inference caused due to the erroneous nature of wireless medium. Route maintenance is initiated when a route failure happens between two nodes. The upstream node of failure tries to repair that route and this process is called local repair.

The proposed protocol uses the local repair technique. AODV can broadcast the RREQ message in the network. The RREQ message is re-broadcasted by the other nodes in the network until it can reach the destination. The route is searched for two times by default in the previous AODV model but in EAODVLRT protocol it can be maximized to seven times in the particular route. In mobile ad hoc network, the mobility of each nodes can be assumed as random way point mobility model [14] with the static pause time. Due to the dynamic nature of the network, the new node may enter to the network that will set the route to the destination i.e. the number of searches to destination for particular route can be increased by seven times.

In the mobile ad-hoc network there is wireless connection between mobile nodes with limited bandwidth [13]. So if the flooding increases in the network the bandwidth will be spread into the network. Then the bandwidth utilization of the nodes in the wireless network will be increased. Here, our aim is to minimize the Flooding; automatically decreasing bandwidth utilization.

In the EAODVLRT algorithm Flooding can be minimized by the local repair technique.

Local repair is a technique used to repair a broken route locally on the upstream nodes of the link failure to the destination no farther than TTLMNR. To repair the link failure, the upstream node broadcasts RREQ packet after increasing the destination sequence number [1]. The TTL value used in RREQ packet is set by the following values:

TTL = Max (0.5 *NH, TTLMNR) +TTLLA

Where:

TTL= Time to live is the constant value which will limits the life span of network, which is generally being used for increasing the number of intermediate nodes in the thesis we had set as 40.

TTLMNR = the last known hop count from the upstream node of the failure to the destination.

TTLLA = constant values which is assumed to be five.

NH = the number of hops from the upstream nodes of the failure to the source of the currently undelivered packet.

TTL_increment = when a route failure happens, the upstream node increments the destination sequence number by two.

 After the upstream node broadcasts the RREQ packet, it waits for the discovery period to receive RREP packets in response to the RREQ packet. When the destination or an intermediate node that has a fresh route to the destination receives the RREQ packet, a RREP packet will be forwarded towards the upstream node. If discovery period finishes and the upstream node didn't receive a RREP for that destination, it transmits a RERR message for that destination to the source. On the other hand, if the upstream node receives one or more RREP packets during the discovery period, it first compares the hop count of the new route with the value in the hop count field of the invalid route table entry for that destination. In the case of the hop count of the newly determined route to the destination is greater than the hop count of the previously known route, the upstream node transmits a RERR message for that destination towards the source, with 'N' bit set. Finally, the upstream node updates its route table entry for that destination.

In Fig. 2 shows the generalized work flow diagram of EAODVLRT which will show the above steps in the diagrammatical form.

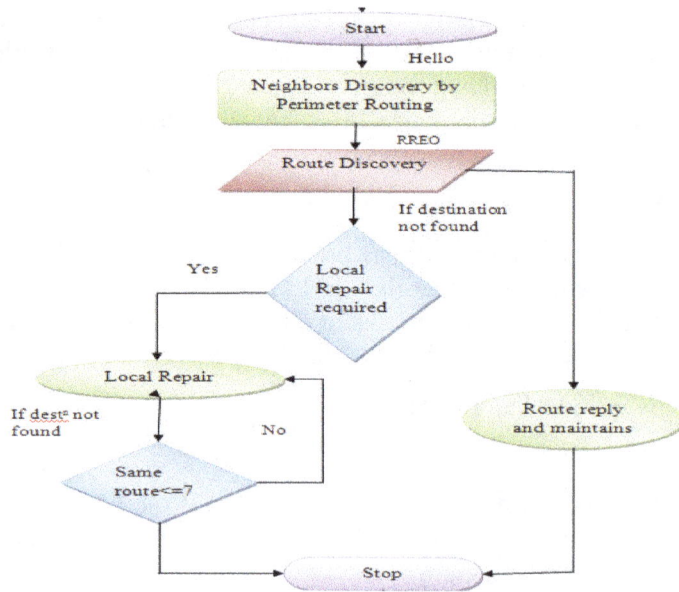

Fig.2 Generalized work flow diagram of EAODVLRT

In AODV TTL_Threshold value should be static in nature. But, in ad hoc network each and every node is dynamic in nature, so every time the topology changes. Therefore, TTL_Threshold value can change every time. By applying this concept TTL_Threshold value can be dynamic in nature.

Once a network is in perimeter method, it prevents the links between nodes from being overlapped by constructing a planar graph using RNG or GG [21, 22]. In this procedure, a packet is forwarded from a start node F to its destination node D guided by the planar graph in clockwise direction as shown in Fig. 3

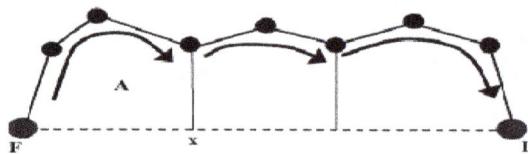

Fig.3 Perimeter forwarding method

In the network perimeter model a packet traverses successively through closer faces of a planar sub graph of the full radio network connectivity graph, until reaching a node closer to the destination.

3.4 Perimeter Algorithm

The main idea behind perimeter forwarding is to forward the packets using the right hand rule across the faces in the planar graph that intercept the line L_pD (Fig.4). The algorithm used for perimeter forwarding [15] is iven below:

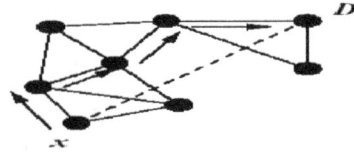

Fig.4 Perimeter Forwarding [8]

The PERIMETER-INIT-FORWARD [23] algorithm forwards a packet p to a node a_{min}, where,

(a_{min}, self) is the first edge encountered countered clockwise from the line L_pD by perimeter routing where p is packet which are send to the destination .

> PERIMETER-INIT-FORWARD (p)
>
> a_{min} = RIGHT-HAND-FORWARD (p, p.D)
>
> return a_{min}

The RIGHT-HAND-FORWARD [16] algorithm implements the right-hand rule method for traversing polygons, which in our case are the faces in the planar graph. The time complexity of the algorithm is O (cN) = O (N), where N is the number of neighbors in the planar graph and c is the time it takes to do a NORM2 operation. NORM can be considered as a constant operation,

Since the range of the arc of the tangent is ($-\pi/2$, $\pi/2$).

> RIGHT-HAND-FORWARD (p, n_{in})
>
> b_{in} = NORM (ATAN2 (self.l.y – n_{in}.y, self.l.x-n_{in}.x)
>
> δ_{min} = 3π
>
> for each (a, l) in N do
>
> > if a == n_{in} then continue
> >
> > b_a = NORM (ATAN2 (self.l.y – l.y, self.l.x – l.x))
> >
> > δ_b = NORM (ba-b_{in})
> >
> > if δ_b<δ_{min} then
> >
> > δ_{min} = δ_b
> >
> > a_{min} = a
> >
> > return a_{min}

If the next edge (self.a, t) in the network is found counter clockwise direction by RIGHT-HAND-FORWARD intercepts L_pD, EAODVLRT updates the packet e_0 field and instead of selecting node t to forward the packet to, it selects the next edge counter clockwise from (self.a, t). We will compare the performance of AODVLRT with the proposed protocol (EAODVLRT) under such conditions for parameters namely throughput, average end to end delay, packet loss, packet delivery fraction.

4   SIMULATION MODEL AND RESULTS

In this section, a series of simulation experiments in NS2 [17] network simulator will be conducted to perform an evaluation analysis on the performance ability of EAODVLRT with the discussed mechanism. We choose ad hoc on demand distance vector (AODV) routing algorithm as the underlying protocol for our base case simulation. AODV is a source initiated reactive (on-demand) protocol, which initiates a route discovery wherever a node requires a path to a destination. Following matrix is used to analyses the performance of EAODVLRT algorithm.

4.1 Simulated Network Scenario and Model

We consider a network of nodes placing in various arrangements within a 1500×600m area. The performance of AODV and EAODVLRT is evaluated by keeping the network size (number of mobile nodes) constant and varying the maximum speed of the nodes. The values of AODVLRT have been taken from base paper [1] which we have implemented. Table 1 show the simulation parameter used in the evaluation.

TABLE 1 Simulation Environment

Dimension of simulated area	1500×600m
Simulation Time	300 sec
Mobile Nodes	50
Transferring Mode	Unicast
Pause time	0,50, 100, 150, 200, 250,300 (m/s)
Traffic	CBR
Packet Size	1024 bytes
Routing Protocols	AODV, Enhanced AODVLR
Transport Layer agent type	TCP
Maximum speed	35
Transmission range	250ms
Mobility	Random
Bandwidth	1 megabits/sec

4.2 Performance Metrics

Analysis of routing protocols for parameters like packet delivery fraction, average end to end delay, packet loss, routing overhead will be done. The parameters are defined in the following section.

- **Packet Delivery Ratio:** It is the ratio of data packets delivered to the destinations to those generated by the CBR sources is known as packet delivery ratio.

Packet Delivery Ratio = packet received / delivered packets

- **Average End To End Delay:** Average end-to-end delay is delay of data packets. Buffering during route discovery latency, queuing at interface queue, retransmission delays at the MAC and transfer times may cause this delay. Once the time difference between every CBR packets sent and received was recorded, dividing the total time difference over the total number of CBR packets received gave the average end-to-end delay for the received packets. Lower the end to end delay better is the performance of the protocol.

Average End to End Delay= Total end to end delay/number of packets received

- **Routing Overhead:** Routing overhead is the number of routing packets transmitted per data packet delivered at the destination. Each hop-wise transmission of a routing packet is counted as one transmission. The first two metrics are the most important for best-effort traffic. The routing load metric evaluates the efficiency of the routing protocol.

Routing Overhead= Total routing packets / transmitted data packets.

- **Average Throughput:** The average number of packets received per amount of time (from the first packet sent to the last packet received).

Average Throughput= Total received packets / simulation time

- **Packet Loss:** It is defined as the difference between the number of packets sent by the source and received by the sink. The routing protocol forwards the packet to destination if a valid route is known; otherwise it is buffered until a route is available. There are two cases when a packet is dropped: the buffer is full when the packet needs to be buffered and the time exceeds the limit when packet has been buffered. Lower is the packet loss better is the performance of the protocol.

Packet Loss = sent packets- received packets.

Note, however, that these metrics are not completely independent. For example, lower packet delivery fraction means that the delay metric is evaluated with fewer samples. In the conventional wisdom, the longer the path lengths, the higher the probability of a packet drops. Thus, with a lower delivery fraction, samples are usually biased in favor of smaller path lengths and thus have less delay.

4.3 Simulation Results and Technical Analysis

EAODVLRT simulation is based on the same environment as for AODV simulation in NS2. Because mobility is the key reason for packet losses, we design the scenarios for comparing the performance of AODV and EAODVLRT based on random mobility model. As indicated in Section 4.A, the simulation environment can be determined by maximum movement speed and the pause time during simulation. This section presents the simulation results and their analysis for 50 nodes network simulation scenario on a rectangular area 1500*600 m^2.

Routing message overhead

The routing message overhead resulted from AODV, AODVLRT [5] and EAODVLRT routing protocols has been presented in Fig. 5. From Fig (5), it could be noticed that EAODVLRT has lower routing message overhead by on average 42% less that the AODV which had been calculated from TABLE 2. For the comparisons purpose values of AODVLRT has being taken from [5]. This result demonstrates the effect of local repair trial by using perimeter routing in EAODVLRT on reducing routing message overhead. This is due to the fact that when the node mobility is increased, the frequency of topology changes is also increased. This can potentially trigger more new route maintenance processes, resulting from the broken routes. As a consequence, larger numbers of RREQ packets are generated and disseminated.

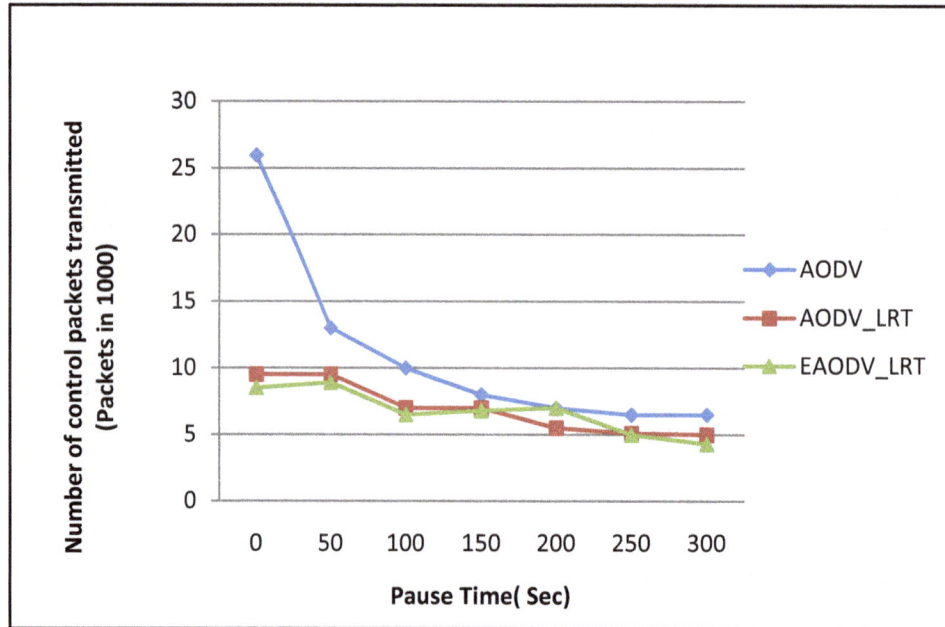

Fig.5 Routing Overhead Analysis

This result demonstrates the effect of local repair trial using perimeter routing in EAODVLRT on reducing routing message overhead. This is due to the fact that when the node mobility is increased, the frequency of topology changes is also increased. As a consequence, larger numbers of RREQ packets are generated and disseminated. Result shows our EAODVLRT routing overhead is minimize as compare to AODV routing protocol. This performance behavior is expected since increasing the offered load leads to an increase in the number of source nodes that initiate route discovery operations.

Throughput

The throughput resulted from AODV, AODVLRT and EAODVLRT has been presented in Fig 6. It can be found that EAODVLRT has higher throughput than AODV routing protocol by an average 1.66% which is a small increase which had being calculated from TABLE 3. It can be found that EAODVLRT has higher throughput than AODV routing protocol by an average 1.66% which is a small increase. This result demonstrates that the effect of the modifications in EAODVLRT doesn't appear in small sized networks. The number of packets dropped or left wait for a route affect the throughput as the increase in the number of packets dropped or left wait for a route reduce the throughput. The number of packets dropped or left wait for a route reduce the throughput. The numbers of packets dropped or left wait for a route affected by the success of local repair in repairing a failed route, where the number of packets dropped or left wait reduced as the percentage of success local repair attempts increased.

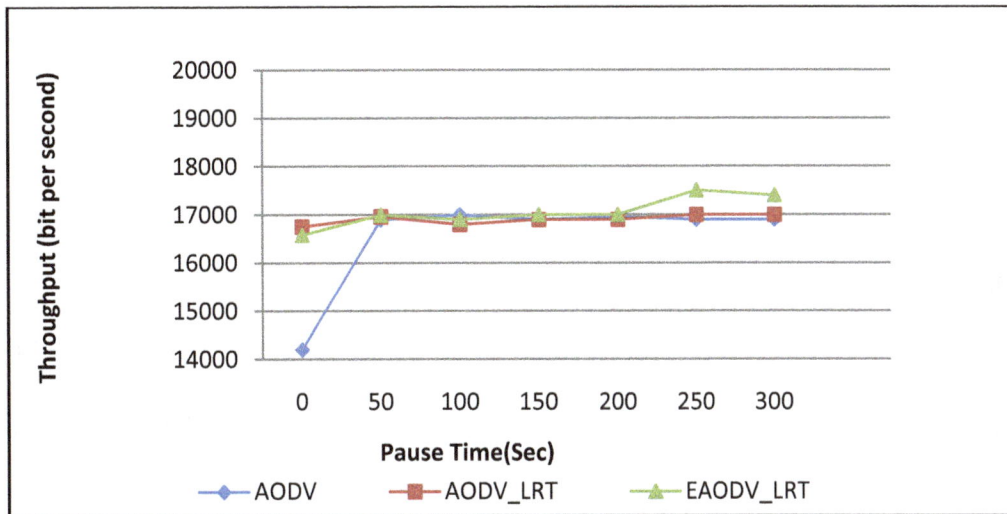

Fig.6 Throughput Analysis

This result returns to that local repair in EAODVLRT acts in trials by broadcasting first RREQ packet with TTL= LR_TTL_START (equal to 2 in experiment). This reduces the routing overhead which by its turn resulted in increasing throughput. On the other side, local repair in AODV broadcasts RREQ packet once with TTL as in AODV local repair in AODV which resulted in higher routing message overhead which led by its turn to reduce the throughput.

Average End to End Delay

Fig 7 presents a graph of packet delivery ratio of AODV and EAODVLRT routing protocols. It is clear that EAODVLRT gives average packet delivery ratio is higher than the AODV by 71.98% which had been calculated from TABLE 4. It is clear that EAODVLRT gives average end to end delay higher than the AODV by on average 28% with difference. The result demonstrates the high effect of local repair trial in EAODVLRT on the delay of the small size of networks which resulted from broadcasting RREQ with TTL as in local repairs. The figure shows that when nodes pause time increases, the end-to-end delay of data packets also increases. This is because the paths between sources and required destinations frequently changed and established. However, among all maximum pause time the AODV performs better, followed by EAODVLRT and AODVLRT.

The increase in the number of broken links will led to increase the delay of transferring packets on a route until finding a repair to the route. The number of broken links affected by the route length as longer routes means the higher chances for broken links. In the same time, the number of broken links affected.

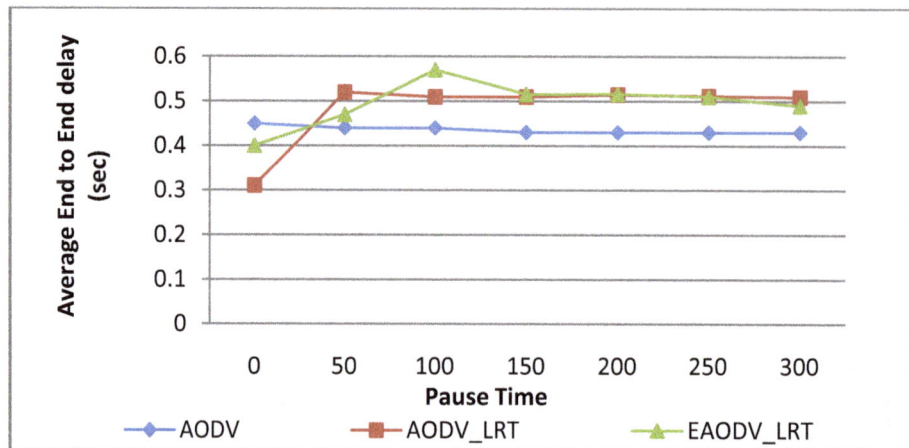

Fig. 7 Average End to End Delay

Packet Delivery Ratio

Fig 8 presents a graph of average end to end delay of AODV, AODVLRT and EAODVLRT routing protocols. It is clear that EAODVLRT gives average end to end delay higher than the AODV by on average 28% with difference, which had being calculated from TABLE 5. It is clear that EAODVLRT gives average packet delivery ratio 71.98%. If we look at this graph, which is for packet delivery ratio of both the protocols, the packet delivery by EAODVLRT is better than AODV. Results shown above concluded that EAODVLRT has packet delivery ratio which is better as compared to AODV Protocol.

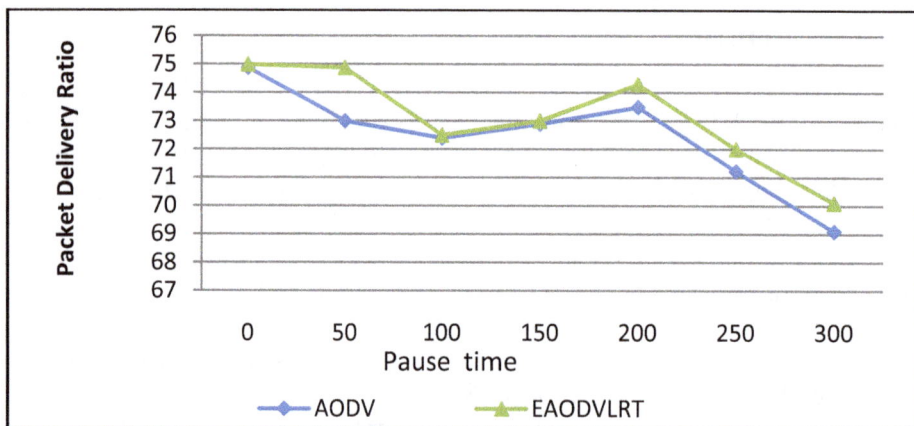

Fig 8 Packet Delivery Ratio

4.3 Overall Summary of AODV and EAODVLRT

The new protocol EAODVLRT perform better than that of AODV routing protocol in terms of "Throughput", "Packet Delivery Ratio", "Routing Overhead" and "End to End Delay". In case of traditional AODV the graph depicts a similarity with EAODVLRT but the values of overall throughput remain less as compared to the EAODVLRT by 1.66%.

Here TABLE 2 shows the comparisons of between AODV and EAODVLRT.

TABLE 2 Comparisons for Throughput

Pause Time	AODV	EAODVLRT
0	14200	16580
50	16900	16990
100	17000	16900
150	16900	17000
200	16955	17000
250	16900	17500
300	16900	17400

TABLE 3 Comparisons for Routing Overhead
(Packets is 1000)

Pause Time	AODV	EAODVLRT
0	26	8.5
50	13	8.9
100	10	6.5
150	8	6.8
200	7	7
250	6.5	5
300	6.5	4.3

Based on Fig. 7 with the variation in pause time, EAODVLRT is performed better in delivering data packet to the destination than AODV. However, when pause time increased, the packet delivery ratio shows the variation on both AODV and EAODVLRT due to the random way point mobility model. Here, TABLE 4 shows the results comparison between both the protocols that is AODV and EAODVLRT.

TABLE 4 Packet Delivery Ratios

Pause Time	AODV	EAODVLRT
0	.45	.4
50	.44	.47
100	.44	.57
150	.43	.515
200	.43	.516
250	.43	.51
300	.43	.49

The Fig. 8 shows that the performance of AODV routing protocol in terms of End to End Delay increases by increasing pause time with random mobility model. EAODVLRT takes less time to deliver the packets. Therefore the optimal delay is achieved but in case of traditional AODV the value of End to End Delay remains less. TABLE 5 shows the average end to end delay of two protocols.

TABLE 5 End to End Delay

Pause Time	AODV	EAODVLRT
0	74.88	74.99
50	73	74.87
100	72.4	72.5
150	72.9	73
200	73.49	74.29
250	71.22	72
300	69.1	70.1

5 CONCLUSIONS

This paper presents a novel approach to minimize routing overheads of AODVLRT. It also analysis enhanced AODVLRT with the existing local repair technique. This approach based on perimeter routing is used to minimize flooding process in EAODVLRT. In this paper we considered the mobile adhoc network routing protocol. Then this work analyzed the issues regarding AODV Local Repair in MANETs while exploring some existing Repair (AODVLRT) technique in literature. This technique consists of three modules. First, broadcasting and can be done by using the perimeter routing. Secondly, flooding is minimized by using local repair method and lastly, number of intermediate nodes from particular source to destination has being increased. This thesis is improved the performance of existing on-demand routing (AODV) protocols by reducing the RREQ overhead during the rout discovery operation. For its implementation and the analysis outcomes NS2 network simulator is used. For analyzing the performance of proposed schema (EAODVLRT) with the existing AODV comparisons had being done. The simulation results show that proposed schema gives the best performance in terms of packet delivery ratio, throughput and number of overhead which is used to compare the performance of these techniques.

REFERENCES:

[1] Maged Salah Eldin Solimana, Sherine Mohamed Abd El-kaderb, Hussein Sherif Eissac, Hoda Anis Barakad "New adaptive routing protocol for MANET" Ubiquitous Computing and Communication Journal, 2006.

[2] C.E Perkins, E. Belding-Royer and S. Das, "Ad Hoc on-Demand Distance Vector (AODV) Routing" RFC 3561, 2003.

[3] Tao Yang, Leonard Barolli, Makoto Ikeda, Fatos Xhafa and ArjanDurresi, "Performance Analysis of OLSR Protocol for Wireless Sensor Networks and Comparison Evaluation with AODV Protocol" IEEE, 2009.

[4] Azzuhri S.R., Portmann M. and Wee Lum Tan "Evaluation of parameterized route repair in AODV" Signal Processing and Communication Systems (ICSPCS), 2010, 4th International Conference on13-15 Dec. 2010.

[5] Mznah A. Al-Rodhaan & Abdullah Al-Dhelaan, "Efficient Route Discovery Algorithm for MANETs" Fifth IEEE International Conference on Networking, Architecture and Storage, 2010.

[6] Joo-Sang Youn, Ji-Hoon Lee, Doo-Hyun Sung, "Quick Local Repair Scheme using Adaptive Promiscuous Mode in Mobile Ad Hoc Networks" journal of networks, vol. 1, no. 1, may 2006.

[7] Kyung-Bae Chang, Dong-Wha Kim, and Gwi-Tae Park "Routing Algorithm Using GPSR and Fuzzy Membership for Wireless Sensor Networks" Springer-Verlag Berlin Heidelberg 2006, pp. 1314 – 1319, 2006.

[8] Brad Karp and H. T. Kung "GPSR: Greedy Perimeter Stateless Routing for Wireless Networks" Mobi Com 2000.

[9] G. Toussaint, "The Relative Neighborhood Graph of a Finite Planar Set Pattern Recognition" Vol. 12, No 4 (1980).

[10] K.Gabriel & R. Sokal "A New Statistical Approach to Geographic Variation Analysis, Systematic Zoology", Vol. 18 (1969), pp. 259–278.

[11] M. Jiang & Y. C. Tay, "Cluster based routing protocol", IETF Internet Draft, draft-ietf-manet-cbrp-spec-02.txt, 1999.

[12] Jagpreet Singh, Paramjeet Singh & Shaveta Ran,"Enhanced Local Repair AODV (ELRAODV)" International Conference on Advances In Computing, Control, and Telecommunication Technologies, IEEE 2009.

[13] Mohd Izuan, Mohd Saad, Zuriati Ahmad Zukarnain, "Performance Analysis of Random-Based Mobility Models in MANET Routing Protocol", European Journal of Scientific Research ISSN 1450-216X Vol.32 No.4 (2009), pp.444-454.

[14] Ching-Wen Chen, Chuan-Chi Weng, "Bandwidth-based routing protocols in mobile ad hoc Networks" Springer Science+Business Media, LLC 2008, pp 240–268.

[15] Maria del Mar Alvarez-Rohena & Chris Eberz, "Implementation and Analysis of GPSR: Greedy Perimeter Stateless Routing for Wireless Networks", IEEE, 2010.

[16] Karp, B., "Geographic Routing for Wireless Networks", Ph.D. Dissertation, Harvard University, Cambridge, MA, October, 2000.

[17] NS2 manualhttp://www.isi.edu/nsnam/ns/doc/ns_doc.pdf.

[18] Liu Chao and Hu Aiqun, "Reducing the Message Overhead of AODV by Using Link Availability Prediction" Springer-Verlag Berlin Heidelberg, LNCS 4864, pp. 113–122, 2007.

[19] L. Nithyanandan, G. Sivarajesh and P. Dananjayan "Modified GPSR Protocol for Wireless Sensor Networks" IEEE, Vol. 2, No. 2, April, 2010.

[20] Tran, D.A. Harish Raghavendra "Congestion Adaptive Routing in Mobile Ad Hoc Networks" Parallel and Distributed Systems, IEEE Transactions onNov. 2006, pp 1294 – 1305.

[21] Jagpreet Singh, Paramjeet Singh &Shaveta Rani, "Enhanced Local Repair AODV (ELRAODV)" IEEE 2009.

3

ADAPTIVE MULTIRESOLUTION MODULATION FOR MULTIMEDIA TRAFFIC OVER NAKAGAMI FADING CHANNELS

Justin James, Annamalai Annamalai, Olusegun Odejide, and Dhadesugoor Vaman

ARO Center for Battlefield Communications, Department of Electrical and Computer Engineering, Prairie View A&M University, Texas 77446
[jjames3,aaannamalai,olodejide,and drvaman]@pvamu.edu

ABSTRACT

Future wireless communications networks will be required to support a multitude of services with different reliability and latency requirements. However, achieving such flexible systems are challenging in Mobile Ad-hoc Networks (MANET) due to limited battery resources, node mobility, changes in path attenuation, and fluctuations in interference. By employing adaptive signaling techniques with Unequal Error Protection (UEP) to take advantage of the differences in the quality of service (QoS) requirements among different types of multimedia traffic, energy efficiency can be improved and also increase the data rate (spectral efficiency) over a fading channel, especially at low and moderate Carrier-to-Noise Ratio (CNR). This paper proposed a combination of curve fitting techniques that derives an exponential approximation to the average packet error rate (PER) probabilities as well as individual PER probabilities for asymmetrical phase-shift keying (PSK) and asymmetrical Quadrature Amplitude Modulation (QAM). Our design is both simpler and more robust than existing schemes.

KEYWORDS

Multiresolution Modulation, Adaptive Modulation, Mobile Ad Hoc Networks, Heterogeneous Traffic, Unequal Error Protection

1. INTRODUCTION

The quality of wireless links in an ad-hoc network is highly variable due to the mobility of radios, fluctuations in path attenuation, transmission impairments due to induced interference, and limited battery resources. Fixed transmission techniques are designed to accommodate worst case link margin channel conditions, therefore, results in suboptimal resource utilization. However, if the channel fade level is known by the transmitter shown in Fig. 1, the Shannon capacity can be approached by adapting the transmit power, data rate, and/or error coding relative to the channel fade level [1-2]. Researchers have shown SNR gains of up to 17 dB by using adaptive modulation techniques compared to that of using non-adaptive modulation in fading environments [3]. The objective of these adaptive signaling techniques is to enhance throughput by transmitting at an elevated data rate when the channel conditions are favorable.

Nearly all of the previously proposed adaptive signaling techniques are intended for systems wherein all of the data to be transmitted has the same quality of service (QOS) requirements. However, future generations of wireless communications networks will be required to support a multitude of services (voice, graphics, video, and data) with a wide variety of reliability requirements and data rates. Adaptive signaling techniques can be designed to exploit the differences in the quality of service requirements among dissimilar types of multimedia traffic.

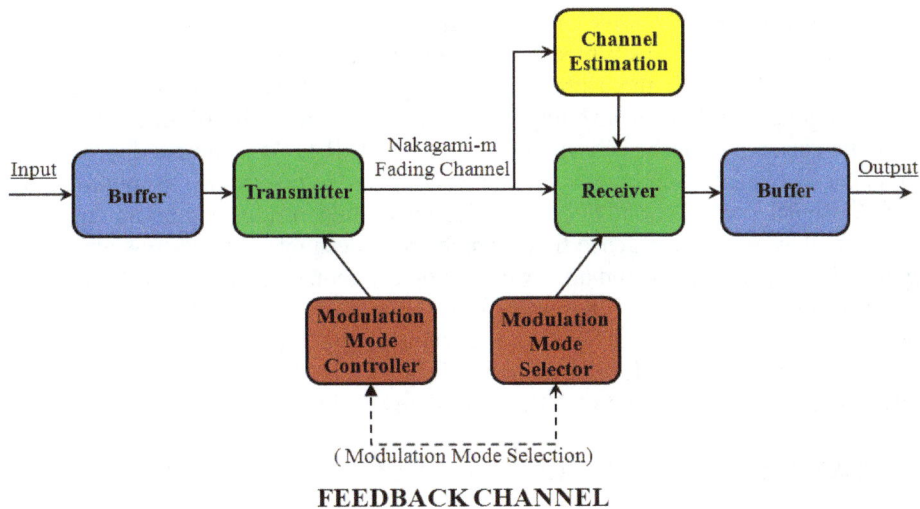

Figure 1. System Model

Recently, substantial research efforts have been dedicated to the integration of voice and data [4-5]. One study proposed adaptive hybrid binary phase shift keying (BPSK)/M-ary Amplitude Modulation (M-AM) as a solution to generate high probability of QoS [4]. This approach assigns the Quadrature (Q) channel to voice with variable power BPSK and the In-phase (I) channel to data with variable-rate M-AM. To exploit the time-varying nature of the channel, power is dynamically allocated between the I and Q channels. The power allocated to the Q channel is just enough to satisfy the target BER for voice, BER_v, and the remaining power is allocated to the I channel to support data with M-AM below the target BER_d. Since M-ary quadrature amplitude modulation (M-QAM) is more spectrally efficient than the M-AM, an adaptive technique utilizing variable-rate uniform M-QAM was proposed in [5] for simultaneous transmission of voice and a single-class of data. However, due to the employment of uniform M-QAM signal constellations, the alphabet size is chosen such that, on average, both the voice and data are transmitted below the target BER for data, BER_d. Consequently, the voice receives unnecessary additional protection at the expense of the spectral efficiency (at low CNR) and outage probability for data.

The concept of transmitting two different classes of data (a basic message and an additional message) using an adaptive non-uniform PSK was proposed in [6]. In [6], both the size and shape (location of the message symbols) of the PSK constellations were proposed to adjust according to the CNR estimates. For this case, the non-uniform M-PSK constellations included two parts: coarse and fine. The coarse part was uniform M-PSK of alphabet size N= M/2. Consequently, the BER of the coarse part is always approximately the same. The fine part consist of the least significant bit (LSB) in the symbol and can have a different probability of BER. One parameter θ, the phase-offset ratio, was proposed to change according to the CNR estimate to increase the throughput for the additional message. In [7], the authors capitalize on the concept of generalized hierarchical (multi-resolution) QAM constellations and their BER expressions to propose a new technique which transmits voice and two different types of data below their respective target BERs. To exploit the time-varying nature of fading, the approach adapts not only the alphabet size but also the priority parameters (a set of parameters, instead of a single parameter as in [6]) of the hierarchical signal constellation, which control the relative BERs of different bit positions. Employing hierarchical M-QAM, the voice bit is transmitted in the LSB-position sub-channel (the least priority sub-channel) of the Q channel. The remaining ($\log_2 M - 1$) sub-channels within the constellation are used for data transmission and allocated to different classes based on the selected priority parameters.

Here, we consider a system in which multiple messages with different QOS requirements are transmitted simultaneously. Each block transmission may include one or more messages in each QOS class. In [6-7], the authors have propose possible solutions to this problem. However, both proposed designs have implementation deficiencies. In [6], the embedded gain is limited by the coarse part of the constellation since the BER for it is always approximately the same. In [7], a sophisticated optimization engine is required to determine the "optimal" shape of the M-QAM signal constellation. Such optimizations could results in non-unique / non-real solutions as well as long convergence times which are undesirable for real-time applications. In this paper, using curve fitting techniques we developed exponential approximation for both the exact average packet error rate (PER) probabilities as well as individual PER probabilities for asymmetrical phase-shift key (PSK) and asymmetrical quadrature amplitude (QAM) modulations. Also, we have developed a simple algorithm for determining the "optimum" phase-offset ratio when more than two UEP classes are considered.

2. ADAPTIVE MULTIRESOLUTION MODULATION WITH UEP

2.1. Background

Previous research has established that the optimal broadcast/multicast scenarios are multiresolution in character due to the competitive nature of broadcast signaling. In multicast environments, the maximum information rate to one user is constrained by the rate of information sent to the other users [8]. However, the proposed conception of interlacing the "coarse" information within the "fine" information is broad in interpretation, and provides no boundaries on the domain in which embedding should be performed. Many researchers have since applied asymmetric signal constellation designs for digital speech transmission over mobile radio channels [9]-[11] and in terrestrial digital video broadcasting [12]-[15] under a joint source-channel coding framework. Others have used non-uniform 4-PSK to decrease the decoder complexity of trellis coded modulation [16]. Additional studies have also investigated the prospect for multicasting multimedia messages to numerous receivers with different capabilities [6], [17], and [18].

Contrary to all previous studies [8]-[18] (excluding [11], [15] and [23]) which are proficient in management of only two levels of UEP despite of the signal constellation size (which restricts the degrees of freedom as well as the data embedding gain), we have created a straightforward algorithm in [11] and [15] for establishing the "optimum" phase-offset ratio when more than two UEP classes are considered. Therefore, it is noteworthy that determining the "optimum" phase-offset ratio is not a trivial undertaking. This is particularly true when more than two UEP levels are employed. Furthermore, [17]-[21] have mainly concentrated on the signal design, predominantly disregarding cross-layer interactions, even though multicast communication also demands interaction with the upper layer protocols. In [22]-[23], we investigated the dilemma of designing a "matched" multi-resolution M-PSK modulation for improving the energy efficiency while concurrently unicasting/multicasting multiple classes of data each requiring a distinct QoS. In [24]-[25], we have also surveyed the effectiveness of multi-resolution modulation in realizing a passive (receiver-oriented) rate-adaptation mechanism without necessitating any feedback from the receiver to the sender for multimedia traffic.

As mentioned in section I, our proposed technique is more robust than [6] and much less complex than [7]. Also, unlike [6] and [7], our design seamlessly switches between M-PSK and M-QAM constellations. To realize our receiver-transmitter oriented rate adaptation strategy using M-PSK and M-QAM modulations, exponential functions were exploited for attaining curve fittings. The proposed receiver-transmitter rate adaptation strategy employs the average PER probability curve fitting approximation to establish which modulation scheme to apply for a specified PER requirement. The individual PER probability curve fitting approximations are then utilized to determine which bit of the asymmetric PSK or QAM constellation to assign to each distinct queue class so that the prescribed PER requirements are satisfied, as in Fig. 3. According to our design, the rate is to be adapted and maximized depending on the prevailing channel conditions (CNR) and the shape is to be adapted based upon the traffic QOS characteristics. In order to perform adaptive modulation, a readily invertible equation in terms of CNR, γ, is needed to determine the constellation size. In order to design an adaptive multiresolution modulation, a readily invertible expression is required to make decisions for both constellation size and shape.

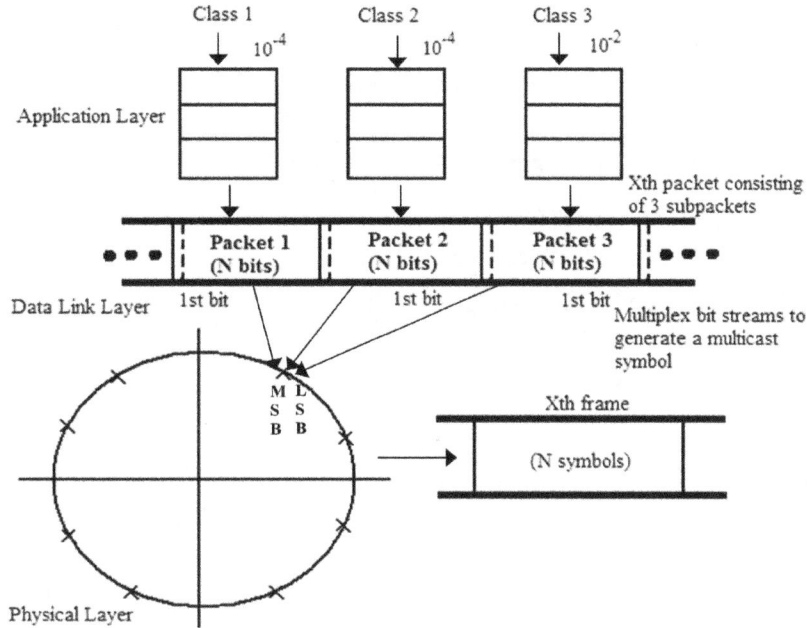

-Leveraging the benefits of assymetric modulation at upper layers of protocol stack

Figure 2. Implementation Design

2.2. System Model

Let's consider, the system model, in Fig. 1, which contains a single-transmitter and single-receiver. The OSI layer implementation of our design is shown in Fig. 2. Our link adaptation strategy employs AMM at the PHY layer to maximize spectral efficiency based on prevailing link conditions and specified QoS requirements. At the application layer, QoS constraints are imposed through target PER requirements. At the physical layer, channel awareness is maintained through channel estimation techniques, such as pilot-based detection). Also, the processing unit at the PHY layer is a frame which is a collection of numerous transmitted symbols.

Seeing that the channel fluctuates from frame-to-frame, we opt to employ the Nakagami-m block channel model to express γ statistically, because it can apply a large class of fading channels with $m = 1$ being equivalent to a Rayleigh fading channel. Additionally useful, there is a direct mapping between the Ricean factor K and the Nakagami fading parameter m. This permits the Nakagami-m channel model to approximate Ricean channels very well. The SNR received per frame, γ, is a random variable with a Gamma probability density function:

$$p_\gamma(\gamma) = \frac{m^m \gamma^{m-1}}{\bar{\gamma}^m \Gamma(m)} \exp\left(-\frac{m\gamma}{\bar{\gamma}}\right) \qquad (1)$$

where $\bar{\gamma} := E\{\gamma\}$ is the average received SNR,

$\Gamma(m) := \int_0^\infty t^{m-1} e^{-t} \, dt$ is the Gamma function, and m is the Nakagami fading parameter ($m \geq 1/2$).

3. PROPOSED SOLUTION

3.1. Exponential Approximation

By harmonizing the distinctive QoS requirements for multimedia sources to the shape of the multicast modulation, a significant embedding gain can be produced. Surprisingly, however, lack of discussion combining adaptive modulation and asymmetric constellations exists. This is attributed to the exact BER expressions found in [19]-[20] are not easily invertible with respect to CNR, γ.

Assuming bit-errors are uncorrelated, the PER (P_B) can be related to the BER (P_b) through

$$P_B = 1 - (1 - P_b)^{N_p} \tag{2}$$

, where N_p is the number of bits in each packet.

The precise BER equations for bit 1 and 2 of 16-QAM in terms of β and γ_s are shown below in equation 3 and 4 respectively.

 As for example, for 16-QAM :

• 4/16 QAM bit one:

$$P_b^{QAM}\left(16_\gamma^\beta, \beta, i_1\right) = \frac{1}{2}\left[\frac{1}{2} erfc\big((\beta^{-1}+1)\Omega\big) + \frac{1}{2} erfc\big((\beta^{-1}-1)\Omega\big)\right] \tag{3}$$

• 4/16 QAM bit two:

$$P_b^{QAM}\left(16_\gamma^\beta, \beta, i_2\right) = \frac{1}{4}\left[2 erfc(\Omega) - erfc\big((2\beta^{-1}+1)\Omega\big) + erfc\big((2\beta^{-1}-1)\Omega\big)\right] \tag{4}$$

, where $\Omega = \sqrt{\dfrac{\gamma}{2(\beta^{-2}+1)}}$.

β the phase offset ratio is as defined in Fig. 3 for Asymmetric 8-PSK and 4/16/64 QAM.

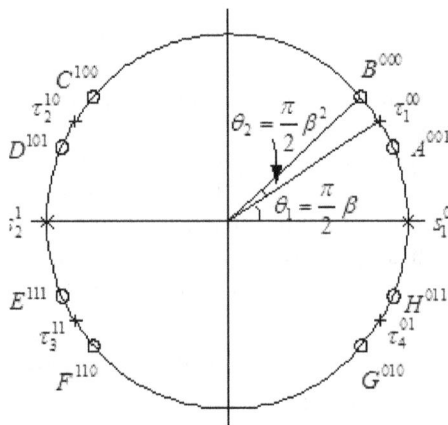

Figure 3a: Asymmetric 8-PSK Constellation

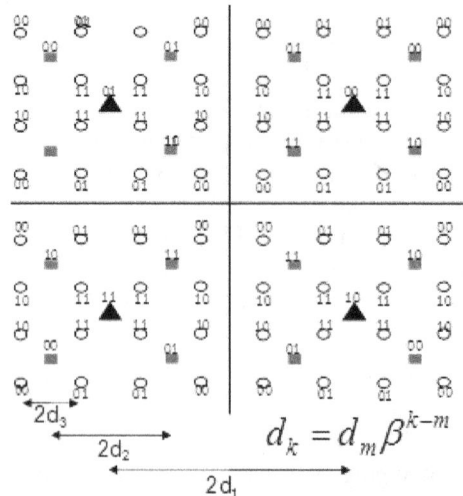

Figure 3b: 4/16/64 QAM Constellation

Figure 3. a) Asymmetric 8-PSK Constellation b) 4/16/64 QAM Constellation

As can be illustrated by equations (3) and (4), the BER expressions are not readily invertible, even for the symmetric case. This difficulty of having a quality readily invertible expression is compounded in the multiresolution modulation case. In this study, this is overcome by approximations developed using the exponential function [25]. Here, we pursue the same exact methods as shown in [25] to develop acceptably accurate approximations. To make this problem analytically tractable, we fix the ratio of angles for any subsequent levels of hierarchy to a constant (viz., $\theta_k = \left(\pi/2\right)\beta^k$) so that only a single design parameter β needs to be optimized as in [6, 25] (instead of manipulating $\log_2 M - 1$ variables) but without sacrificing the ability of higher order alphabets (i.e., denser signal constellations) to support a more flexible multimedia transmission. Our simplified asymmetric PSK and QAM constellations also revert to the uniform PSK case when $\beta = 0.5$. We have also extended our previous efforts using exponential approximation for asymmetrical M-QAM and M-PSK and generated the curve fitting approximations for the average PER probability and individual PER for 4-PSK, 8-PSK, 16-PSK, 16-QAM, 64-QAM, and 256-QAM.

The exponential approximation equations used for M-QAM and M-PSK modulations in terms of β and γ_s are illustrated below in equations 5 and 6.

$$P_B^{(i)}(\gamma_s, \beta) = 1 \qquad\qquad\qquad \text{if} \quad \gamma_s < d_\beta^{(i)} \qquad\qquad (5)$$

$$P_B^{(i)}(\gamma_s, \beta) = a_\beta^{(i)} e^{-b_\beta^{(i)}\gamma_s} + c_\beta^{(i)} e^{-2b_\beta^{(i)}\gamma_s} \qquad \text{if} \quad \gamma_s \geq d_\beta^{(i)} \qquad\qquad (6)$$

- where i = 1, 2, 3..., $\log_2 M$, β is the phase-offset ratio, γ_s is the average CNR, and $d_\beta^{(i)} = \gamma_{THRESHOLD}$ which denotes the knee value of the plot.

- $a_\beta^{(i)}$, $b_\beta^{(i)}$, and $c_\beta^{(i)}$ are expressed in 3rd order polynomial of β, $p_3\beta^3 + p_2\beta^2 + p_1\beta + p_0$.

- p_4, p_3, p_2, p_1, and p_0 are constants to be determined from the exponential approximation.

- $d_\beta^{(i)}$ is expressed by a 4th order polynomial of β.

$a_\beta^{(i)}$, $b_\beta^{(i)}$, $c_\beta^{(i)}$, and $d_\beta^{(i)}$ are four parameters to be resolved such that the disparity between the exact PER and the estimate is minimized in the sense of mean square error. For this paper, we employed the Quasi-Newton BFGS algorithm to execute the curve fitting. The error tolerance of the utilized curve fitting algorithm was set to be $1e^{-10}$.

3.2. Curve Fitting Results / Applications

Using the exponential function, approximations were developed. For the exact curve, the complicated exact PER formulas were applied for ($\{0.1 - 0.5\}$ $\beta \in$) in an AWGN channel. Curve fitting results in Fig. 4 for exact and approximate PER probability (individual bits) for 16-QAM are illustrated. Curve fitting results for the exact and approximate average PER probability for different multi-resolution modulation schemes are illustrated in Fig. 5. From Fig. 4 and Fig.5, it can be shown that our invertible exponential expression yields a very tight approximation of the exact curves. Also, as shown in Fig. 4, as β deviates from 0.5, the QOS disparity between $P_B^1(\gamma_s)$ and $P_B^2(\gamma_s)$ clearly increases. Consequently, classical symmetric modulation is well-suited for homogeneous traffic transmission where all bits have relatively the same QoS requirements. However, for heterogeneous traffic and multimedia applications where individual bits can have different QoS requirements, multiresolution modulation can significantly enhance spectral efficiency and system performance.

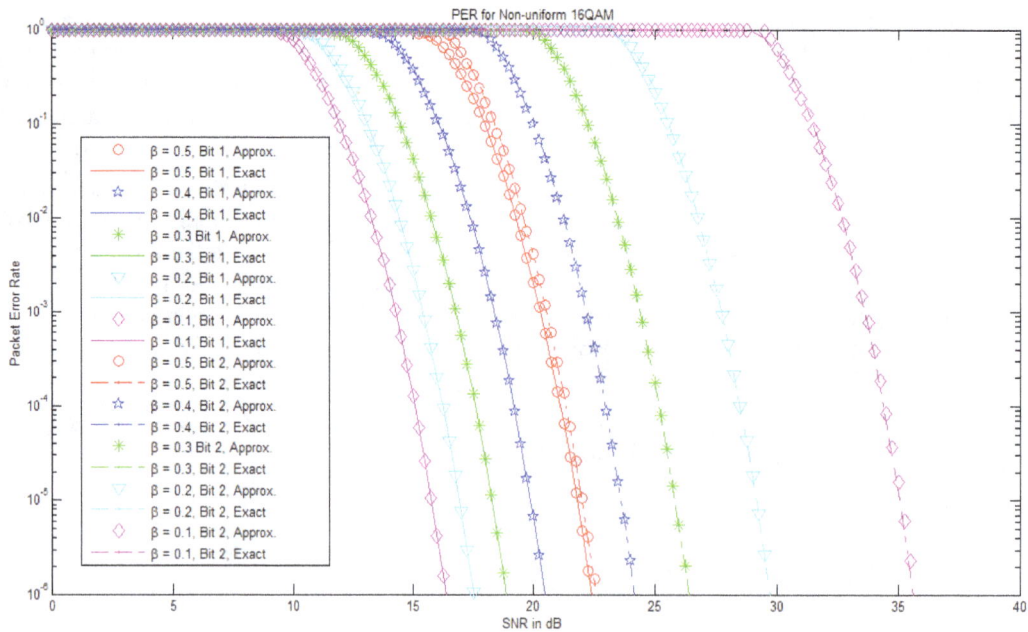

Figure 4. Exact and approximate PER for 16-QAM (individual bit class)

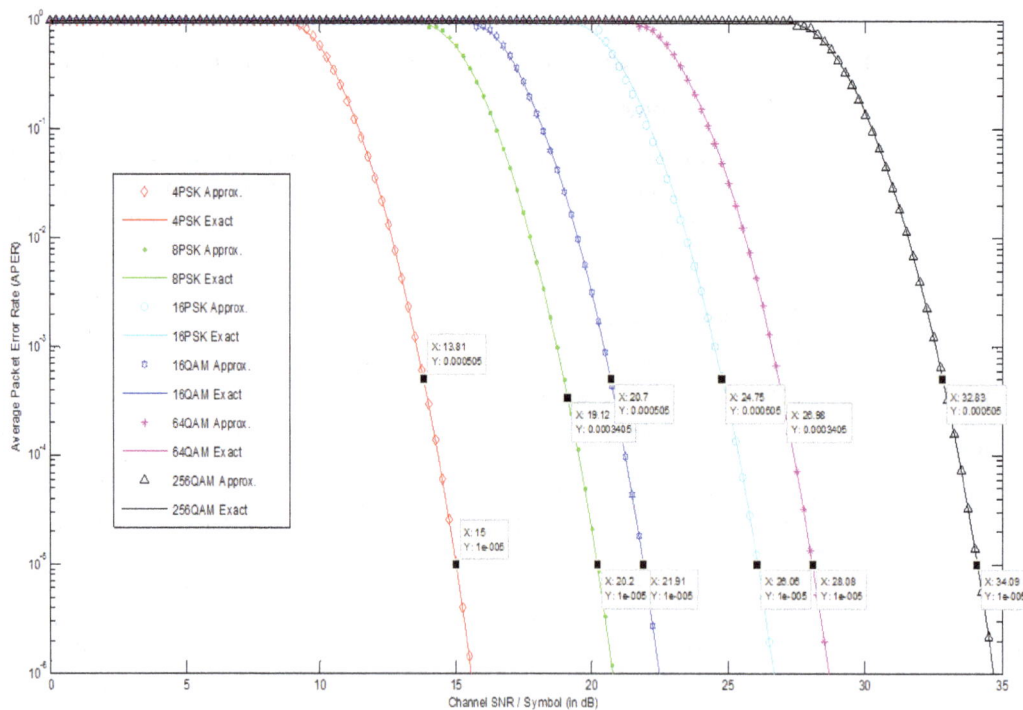

Figure 5: Exact and approximate average PER probability and the threshold values for both conventional and adaptive multiresolution schemes given the specified QOSs 10^{-5} and 10^{-3}.

The polynomial coefficients (p_4, p_3, p_2, p_1, and p_0) for these figures are in table 1 and 2. We also created curve fittings for 4-PSK, 8-PSK, 16-PSK, 64-QAM, and 256-QAM. A summary of polynomial coefficients for the average and individual PER probabilities using the exponential approximation for 4-PSK, 8-PSK, 16-PSK, 64-QAM, and 256-QAM are available through the appendix (Packet size = 1080 bits/packet).

Table 1. POLYNOMIAL COEFFICIENTS FOR 16-QAM (INDIVIDUAL BIT CLASS) PROBABILITY

		P_4	P_3	P_2	P_1	P_0
1	$a_\beta^{(1)}$	N/a	-464.5364	472.3193	-151.9469	83.8828
	$b_\beta^{(1)}$	N/a	0.1675	0.5434	-1.1648	0.5296
	$c_\beta^{(1)}$	N/a	2.0579e+04	-2.0527e+04	6.4391e+03	-1.9225e+03
	$d_\beta^{(1)}$	38.77	-59.33	45.17	-0.1344	8.955
2	$a_\beta^{(2)}$	N/a	567.4087	-1.0637e+03	498.5439	54.5480
	$b_\beta^{(2)}$	N/a	-0.4291	0.6580	-0.0196	7.4264e-04
	$c_\beta^{(2)}$	N/a	-7.2657e+04	1.1728e+05	-5.1231e+04	2.8093e+03
	$d_\beta^{(2)}$	328.2000	-553.8000	374.9000	-136.9000	39.4100

Table 2. POLYNOMIAL COEFFICIENTS FOR 16-QAM (AVERAGE BIT CLASS) PROBABILITY

	P_4	P_3	P_2	P_1	P_0
a_β	N/a	0	0	0	101.4961
b_β	N/a	0	0	0	0.1038
c_β	N/a	0	0	0	-2.9007e+03
d_β	649.4000	-908.0000	514.1000	-159.7000	39.9400

Let's consider a circumstance whereby the channel necessitates maintaining two different types of traffic demanding two different QoS's. The QoS's respectively are 10^{-4} and 10^{-2}. This is illustrated in Table 3.

Table 3. QOS demands for different modulation types a) QOSs 10^{-5} & 10^{-3} b) QOSs 10^{-4} & 10^{-2}

***To find the CNR_{REQ}, the QOS demands must be known for PER_{AVG}.		
For (4PSK, 16QAM)	For (8PSK, 64QAM)	For ((16PSK, 256QAM)
$P_B^{(1)} \leq 10^{-5}$ $P_B^{(2)} \leq 10^{-3}$ $\beta \in [0,0.5]$	$P_B^{(1)} \leq 10^{-5}$ $P_B^{(2)} \leq 10^{-5}$ $P_B^{(3)} \leq 10^{-3}$ $\beta \in [0,0.5]$	$P_B^{(1)} \leq 10^{-5}$ $P_B^{(2)} \leq 10^{-5}$ $P_B^{(3)} \leq 10^{-3}$ $P_B^{(4)} \leq 10^{-3}$ $\beta \in [0,0.5]$
$\dfrac{10^{-5} + 10^{-3}}{2}$ $= 0.000505$	$\dfrac{10^{-5} + 10^{-5} + 10^{-3}}{3}$ $= 0.00034$	$\dfrac{10^{-5} + 10^{-5} + 10^{-3} + 10^{-3}}{4}$ $= 0.000505$

(a)

***To find the CNR_{REQ}, the QOS demands must be known for PER_{AVG}.		
For (4PSK, 16QAM)	For (8PSK, 64QAM)	For ((16PSK, 256QAM)
$P_B^{(1)} \leq 10^{-4}$ $P_B^{(2)} \leq 10^{-2}$ $\beta \in [0,0.5]$	$P_B^{(1)} \leq 10^{-4}$ $P_B^{(2)} \leq 10^{-4}$ $P_B^{(3)} \leq 10^{-2}$ $\beta \in [0,0.5]$	$P_B^{(1)} \leq 10^{-4}$ $P_B^{(2)} \leq 10^{-4}$ $P_B^{(3)} \leq 10^{-2}$ $P_B^{(4)} \leq 10^{-2}$

		$\beta \in [0,0.5]$
$\dfrac{10^{-4} + 10^{-2}}{2}$ $= 0.00505$	$\dfrac{10^{-4} + 10^{-4} + 10^{-2}}{3}$ $= 0.0034$	$\dfrac{10^{-4} + 10^{-4} + 10^{-2} + 10^{-2}}{4}$ $= 0.00505$

(b)

Depending on the channel condition, the scheme should be able of determining the best modulation scheme to employ in order to meet the required QoS. This can be done either by a labor-intensive method (plotting the curve using the exact equation, and manually establishing the switching threshold values through graphical analysis for different modulation schemes), or by utilizing the invertible expression to forecast the γ_{REQ}. The manual technique is rigorous and cannot be automated. Thus, the invertible expression in eqn. 7, which is the inversion of eqn. 6, supplies an extremely straightforward and simple approach for determining the switching threshold values for each.

$$y_\beta^{(i)} = \frac{\left(-a_\beta^{(i)} + \sqrt{\left(a_\beta^{(i)} \right)^2 + 4c_\beta^{(i)} P_B^{(i)}} \right)}{2c_\beta^{(i)}} \tag{7}$$

$$\gamma_{req}^{(i)}(\beta) = -\frac{1}{b_\beta^{(i)}} \ln y_\beta^{(i)}, \qquad i = 1,2,3, \dots \log_2 M \tag{8}$$

Employing our approach, the curve for the average PER rate is necessary for inter-modulation threshold determination. The required QoS for the average PER is the average of the individual QOS's. For the case where there are two distinct class, the QOS requirements would be $(10^{-4} + 10^{-2}) / 2$. After determining the mandatory CNRs that designates which modulation scheme to exploit for a particular range of channel conditions, we also must establish the CNR threshold values to be exploited within each modulation schemes in order to cater to the QoS requirements for the individual bits. These CNR threshold values are referred to as intra-modulation thresholds. For both the cases (average and individual), the threshold values are listed in Tables 5 & 6 respectively. Within each modulation scheme, we decide to have the most rigorous provision for the first bit and the next condition for the next bit after that. Following this, we use 10^{-4} for bit 1 and 10^{-2} for bit 2, for the case of 4/16 QAM modulation scheme. Conversely, for the usage of 4/16/64 QAM modulation scheme, we opt to have 10^{-4} for bit 1, 10^{-4} for bit 2 and 10^{-2} for bit 3.

3.3. Optimum Phase-Offset Ratio Algorithm

To facilitate determining the optimum β value for each modulation scheme, we should decide on the β value that results in the minimum of the maximum γ_{req}^i.

$$\arg \; \min_{\beta} \; \max_{i} \; \{ \{ \gamma_{req}^i \} \} \tag{9}$$

For the particular case of 4/16/64 QAM, the QoS's are defined as,

- $P_B^{(1)} \leq 10^{-4}$
- $P_B^{(2)} \leq 10^{-4}$
- $P_B^{(3)} \leq 10^{-2}$

$\beta \in (0,0.5], i \in \{1,2,3\}$.

4. COMPUTATIONAL RESULTS

Acquiring the optimum β value for different modulation scheme guarantees that each modulation scheme maximizes its capability based on the specified QoS requirements. It is also worth noting that this is attainable with the aid of the invertible expression found in eqn. 7, which is a direct application of the curve fitting process. The optimum phase-offset ratios, β, for different modulations with respect to QOS demands are in Table 4.

Table 4. Optimum beta value for different modulations a) QOSs 10^{-5} & 10^{-3} b) QOSs 10^{-4} & 10^{-2}

Modulation	Optimum Beta
4PSK	0.450
8PSK	0.485
16PSK	0.490
16QAM	0.465
64QAM	0.465
256QAM	0.490

(a)

Modulation	Optimum Beta
4PSK	0.440
8PSK	0.485
16PSK	0.485
16QAM	0.455
64QAM	0.460
256QAM	0.485

(b)

The assorted CNR threshold, γ_{REQ}, values obtained for inter-modulation schemes are listed in Table 5, while the values obtained for intra-modulation switching thresholds (the CNR values at which the requirements for each individual bits are being accommodated within different modulation schemes) are revealed in Table 6.

Table 5. Threshold values (in dB) for the Inter-modulation scheme a) QOSs 10^{-5} & 10^{-3} b) QOSs 10^{-4} & 10^{-2}

Modulation	4PSK	8PSK	16PSK	16QAM	64QAM	256QAM
Conventional	15.00	20.20	26.06	21.91	28.08	34.09
Adaptive	13.81	19.12	24.75	20.70	26.98	32.83

(a)

Modulation	4PSK	8PSK	16PSK	16QAM	64QAM	256QAM
Conventional	14.34	19.53	25.36	21.24	27.39	33.39
Adaptive	12.93	18.26	23.81	19.80	26.09	31.89

(b)

Table 6. Threshold values (in dB) for the Intra-modulation scheme a) QOSs 10^{-5} & 10^{-3} b) QOSs 10^{-4} & 10^{-2}

Modulation	Bit Class	α=β=0.3	α=β=0.4	α=β=0.5	α=β=optimum	QOS$_{PER_{AVG}}$
4PSK	Bit 1	12.99	13.83	15.00	14.37	10^{-5}
	Bit 2	17.42	15.17	13.57	14.31	10^{-3}
8PSK	Bit 1	13.55	15.71	20.15	19.23	10^{-5}
	Bit 2	21.59	20.48	20.15	20.15	10^{-5}
	Bit 3	27.58	22.67	18.87	19.38	10^{-3}
16PSK	Bit 1	13.66	16.69	25.80	24.11	10^{-5}
	Bit 2	22.55	22.88	25.80	25.32	10^{-5}
	Bit 3	30.40	26.76	24.75	24.67	10^{-3}
	Bit 4	38.01	30.53	24.75	25.28	10^{-3}
16QAM	Bit 1	18.29	19.89	21.80	21.10	10^{-5}
	Bit 2	24.40	22.17	20.56	21.07	10^{-3}
64QAM	Bit 1	19.32	22.49	27.84	25.56	10^{-5}
	Bit 2	28.78	27.95	28.04	27.91	10^{-5}
	Bit 3	34.89	30.22	26.79	27.88	10^{-3}
256QAM	Bit 1	19.53	23.66	33.70	31.94	10^{-5}
	Bit 2	29.78	30.46	33.91	33.37	10^{-5}
	Bit 3	37.73	34.41	32.60	32.72	10^{-3}
	Bit 4	45.35	38.20	32.86	33.33	10^{-3}

(a)

Modulation	Bit Class	α=β=0.3	α=β=0.4	α=β=0.5	α=β=optimum	QOS$_{PER_{AV}}$
4PSK	Bit 1	12.33	13.17	14.34	13.60	10^{-4}
	Bit 2	16.48	14.24	12.63	13.53	10^{-2}
8PSK	Bit 1	12.86	15.02	19.48	18.53	10^{-4}
	Bit 2	20.90	19.79	19.46	19.46	10^{-4}
	Bit 3	26.64	21.71	17.96	18.47	10^{-2}
16PSK	Bit 1	12.96	15.97	25.08	22.67	10^{-4}
	Bit 2	21.83	22.16	25.08	24.37	10^{-4}
	Bit 3	29.40	25.76	23.81	23.76	10^{-2}
	Bit 4	37.07	29.59	23.81	24.60	10^{-2}
16QAM	Bit 1	17.59	19.20	21.11	20.21	10^{-4}
	Bit 2	23.46	21.23	19.62	20.29	10^{-2}
64QAM	Bit 1	18.59	21.76	27.11	24.55	10^{-4}
	Bit 2	28.08	27.26	27.34	27.21	10^{-4}
	Bit 3	33.95	29.29	25.85	27.11	10^{-2}
256QAM	Bit 1	18.78	22.90	32.94	30.41	10^{-4}
	Bit 2	29.05	29.74	33.19	32.39	10^{-4}
	Bit 3	36.73	33.41	31.60	31.79	10^{-2}
	Bit 4	44.41	37.26	31.92	32.64	10^{-2}

(b)

When mode n is employed, each symbol transmitted will contain $R_n = log_2(M_n)$ information bits. We presume a Nyquist pulse shaping filter with bandwidth, $B=1/T_s$, where T_s is the symbol rate. For the adaptive multiresolution scheme, the average spectral efficiency (bit rate per unit bandwidth) can be written as shown in equation (10).

$$
\begin{aligned}
\eta_{adap} = {} & 0 \int_{\gamma_{T1}}^{\gamma_{4p1}} f_\gamma(\gamma)\,\delta_\gamma + 1 \int_{\gamma_{4p1}}^{\gamma_{4p2}} f_\gamma(\gamma)\,\delta_\gamma + 2 \int_{\gamma_{4p2}}^{\gamma_{T2}} f_\gamma(\gamma)\,\delta_\gamma + 0 \int_{\gamma_{T2}}^{\gamma_{8p1}} f_\gamma(\gamma)\,\delta_\gamma \\
& + 1 \int_{\gamma_{8p1}}^{\gamma_{8p2}} f_\gamma(\gamma)\,\delta_\gamma + 2 \int_{\gamma_{8p2}}^{\gamma_{8p3}} f_\gamma(\gamma)\,\delta_\gamma + 3 \int_{\gamma_{8p3}}^{\gamma_{T3}} f_\gamma(\gamma)\,\delta_\gamma + 0 \int_{\gamma_{T3}}^{\gamma_{16Q1}} f_\gamma(\gamma)\,\delta_\gamma \\
& + 2 \int_{\gamma_{16Q1}}^{\gamma_{16Q2}} f_\gamma(\gamma)\,\delta_\gamma + 4 \int_{\gamma_{16Q2}}^{\gamma_{T4}} f_\gamma(\gamma)\,\delta_\gamma + 0 \int_{\gamma_{T4}}^{\gamma_{16p1}} f_\gamma(\gamma)\,\delta_\gamma \\
& + 1 \int_{\gamma_{16p1}}^{\gamma_{16p2}} f_\gamma(\gamma)\,\delta_\gamma + 2 \int_{\gamma_{16p2}}^{\gamma_{16p3}} f_\gamma(\gamma)\,\delta_\gamma + 3 \int_{\gamma_{16p3}}^{\gamma_{16p4}} f_\gamma(\gamma)\,\delta_\gamma \\
& + 4 \int_{\gamma_{16p4}}^{\gamma_{T5}} f_\gamma(\gamma)\,\delta_\gamma + 0 \int_{\gamma_{T5}}^{\gamma_{64Q1}} f_\gamma(\gamma)\,\delta_\gamma + 2 \int_{\gamma_{64Q1}}^{\gamma_{64Q2}} f_\gamma(\gamma)\,\delta_\gamma \\
& + 4 \int_{\gamma_{64Q2}}^{\gamma_{64Q3}} f_\gamma(\gamma)\,\delta_\gamma + 6 \int_{\gamma_{64Q3}}^{\gamma_{T6}} f_\gamma(\gamma)\,\delta_\gamma + 0 \int_{\gamma_{T6}}^{\gamma_{256Q1}} f_\gamma(\gamma)\,\delta_\gamma \\
& + 2 \int_{\gamma_{256Q1}}^{\gamma_{256Q2}} f_\gamma(\gamma)\,\delta_\gamma + 4 \int_{\gamma_{256Q2}}^{\gamma_{256Q3}} f_\gamma(\gamma)\,\delta_\gamma + 6 \int_{\gamma_{256Q3}}^{\gamma_{256Q4}} f_\gamma(\gamma)\,\delta_\gamma \\
& + 8 \int_{\gamma_{256Q4}}^{\infty} f_\gamma(\gamma)\,\delta_\gamma
\end{aligned}
$$

(10)

where γ_{T1}, γ_{T2}, γ_{T3}, γ_{T4}, γ_{T5}, and γ_{T6}, are the adaptive scheme's inter-modulation threshold values for 4PSK, 8PSK, 16QAM, 16PSK, 64QAM, and 256QAM respectively. γ_{4P1}, and γ_{4P2}, are the intra modulation threshold values for 4PSK. In a similar manner, γ_{16Q1} and γ_{16Q2}, are the intra modulation threshold values for 16QAM. Equation (10) is applicable in a situation whereby $\gamma_{T1} < \gamma_{4P1} < \gamma_{4P2} < \gamma_{T2} < \gamma_{8P1} < \gamma_{8P2} < \gamma_{8P3} < \gamma_{T3} < \gamma_{16Q1} < \gamma_{16Q2} < \gamma_{T4} < \gamma_{16P1} < \gamma_{16P2} < \gamma_{16P3} < \gamma_{16P4} < \gamma_{T5} < \gamma_{64Q1} < \gamma_{64Q2} < \gamma_{64Q3} < \gamma_{T6} < \gamma_{256Q1} < \gamma_{256Q2} < \gamma_{256QP3} < \gamma_{256QP4}$.

Figures 6 and 7 show the spectral efficiency using under Nakagami block fading channel conditions, while Fig. 8 and 9 show the spectral efficiency as a function of the Nakagami fading parameter, m, which is equivalent to the order of diversity. The results are shown for different β values including normal transmission in which β = 0.5 and the optimum β value.

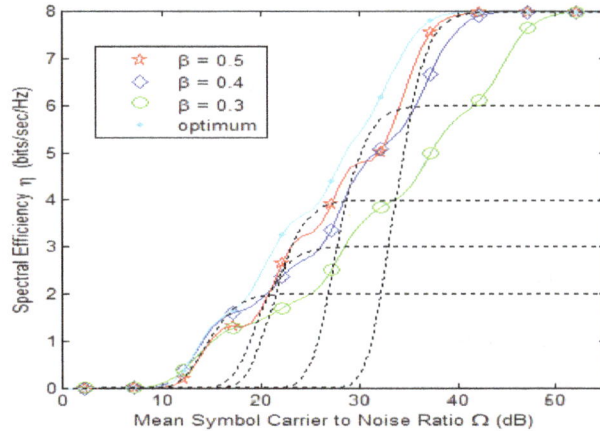

Figure 6. Spectral Efficiency for different beta values using QOSs 10^{-4} & 10^{-2} (nakagami-m=3)

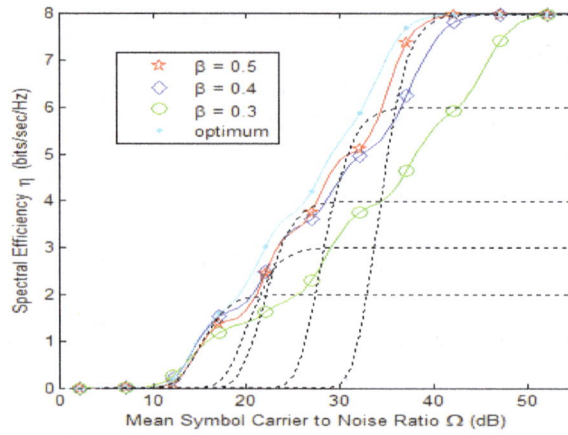

Figure 7. Spectral Efficiency for different beta values using QOSs 10^{-5} & 10^{-3} (nakagami-m=3)

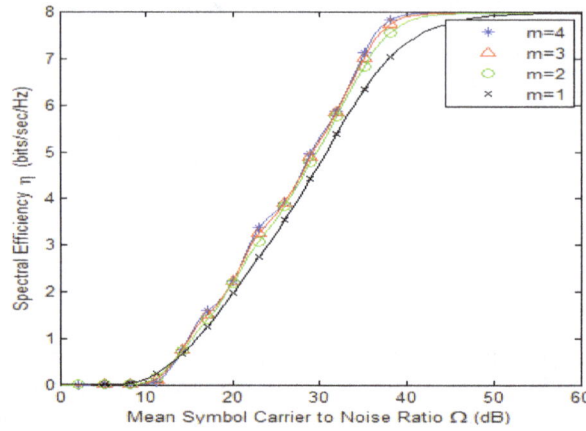

Figure 8. Spectral Efficiency for different nakagami-m fading parameter using QOSs 10^{-5} & 10^{-3}

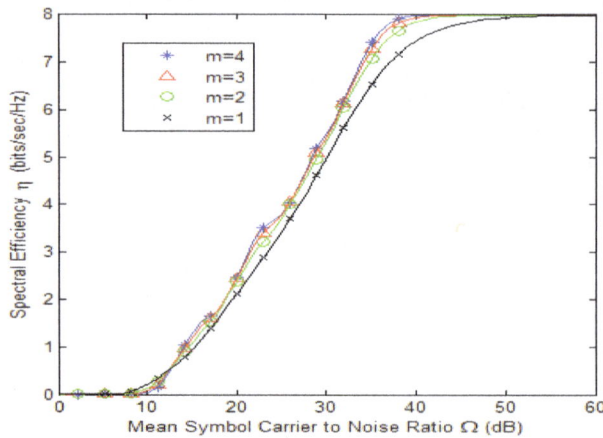

Figure 9. Spectral Efficiency for different nakagami-m fading parameter using QOSs 10^{-4} & 10^{-2}

Figures 10 and 11 show the results obtained for the average probabilty of packet error rate using adaptive multiresolution modulation techniques based on the specified QOSs under Nakagami fading channel conditions. The results were obtained by using the switching threshold values in Table 4 and 5.

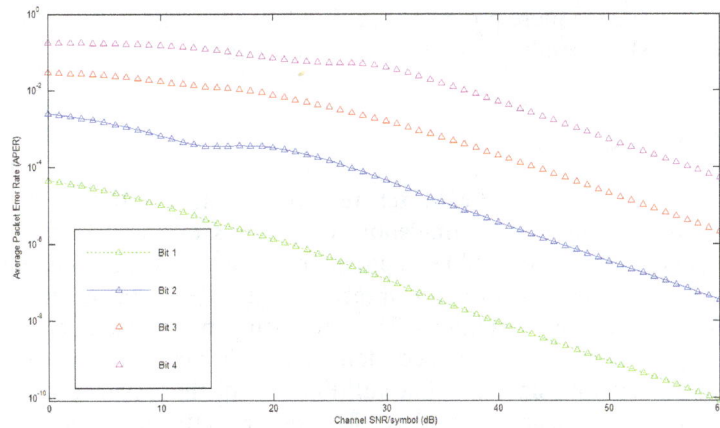

Figure 10. Average PER using m=1, β=0.45, and QOSs 10^{-5} and 10^{-3}

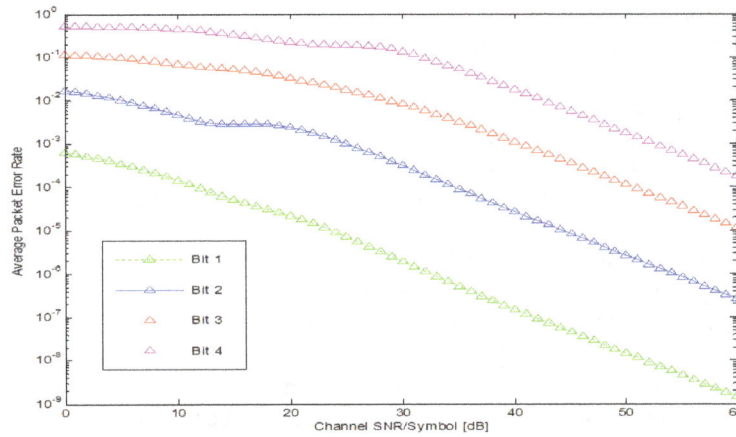

Figure 11. Average PER using m=1, β=0.45, and QOSs 10^{-4} and 10^{-2}

From Tables 5 & 6, it is apparent that our multiresolution modulation technique in comparison to the conventional scheme reduces the (inter-modulation and intra-modulation) switching threshold values for each case. These decreased switching thresholds aid in facilitating higher spectral efficiency.

From fig(s). 6 and 7, we note that, if phase-offset ratio β is fixed, conventional symmetrical modulation (β=0.5) would be best in the high SNR / CNR range. However, for low and moderate SNR / CNR, the multiresolution modulation proposed is more advantageous. This is because of the UEP facilitated through asymmetric modulation. Since, standard symmetric modulations give equal priority to each bit within the symbol, channel conditions affect each bit equally. However, since multiresolution modulation maps more the important bits to higher priority positions within the transmitted symbol, the bits are unequally protected. Thus, in the case that we receive and an erroneous symbol, we can recover the more important bits. As β is reduced, the unequal protection disparity becomes greater amongst the bit classes. Consequently, if β was fixed, β=0.3 would perform best in the low SNR / CNR range, β=0.4 would perform best in the moderate SNR / CNR range, and β=0.5 would perform best in the high SNR / CNR range. However, employing our algorithm (9) for determining the optimum β for the specified QOS's, we acquired the best spectral efficiency across the low. moderate, and high SNR / CNR ranges.

Figures 10 and 11 illustrate the average PER for each bit class based on the prescribed QoS constraints. From figs. 10 and 11, it is obvious that the bit 1 (the most significant bit) is more protected than bit 2 which is more protected than bit 3 which is more protected than bit 4 (the least significant bit). These results illustrate the efficacy gains available through multiresolution modulation.

5. CONCLUSIONS

The advantages of employing an AMM scheme for multimedia traffic in MANET were inspected by means of analysis and simulation. Our results exemplified the significance of having a high-quality readily invertible approximation of the exact equation to facilitate automatically making decisions on switching thresholds for different modulation schemes based on the required QoS and channel conditions. This approach permits adaptation of both the shape and the size of the asymmetric signal constellation with the intention of improving the throughput for all the multimedia data while fulfilling the distinct QoS requirements for each traffic type under different channel conditions. The spectral efficiency gain achieved through the employment of multiresolution modulation at the physical layer is considerable, especially at low and moderate CNR. This performance gain is obtained by mapping higher priority buffer bits to higher protected bit positions within the symbol. In comparison to the previously proposed techniques, our approach is simpler and more robust.

6. FUTURE WORKS

Moving forward, we intend to extend this work to develop a cross-layer design based on QoS demands at the application layer in terms of latency and reliability, an ARQ at the data link layer to allow retransmissions (thus lowering the stringent demands at the PHY layer), and AMM at the physical layer to compensate for link quality fluctuations and to improve the spectral efficiency. Also, we will aim to extend this work by employing joint adaptive multi-resolution modulation with retransmission diversity which captures the disparity in the QOS requirements for dissimilar traffic types. We also plan to evaluate the packet loss probability due to queuing (finite buffer length) by means of numerical analysis and comprehensive MATLAB simulations. Also, diversity combining techniques will be employed to merge multiple copies of received signals at the receiver into a single enhanced signal.

The prime objective of this research is to establish the "preeminent" approach that can concurrently support transmission of guaranteed real-time services and best-effort services over deprived wireless links. The throughput and delay performance of the ARQ protocol will be additionally enhanced by exploiting packet combining schemas at the data link layer. In particular, we will investigate the benefits with simple pre-detection / post-detection combining techniques, such as majority logic decoding and average diversity combiner. Though suboptimal, such approaches can extensively compress the buffer requirements and implementation complexity.

ACKNOWLEDGEMENTS

This research work is supported in part by the U.S. Army Research Laboratory under Cooperative Agreement W911NF-04-2-0054 and the National Science Foundation NSF/HRD 0931679. The views and conclusions contained in this document are those of the authors and should not be interpreted as representing the official policies, either expressed or implied, of the Army Research Laboratory or the National Science Foundation or the U. S. Government.

REFERENCES

[1]   ITU-T Recommendation H.264 Advanced Video Coding for Generic Audiovisual Services, May 2003.

[2]   A. J. Goldsmith and P. Varaiya, "Capacity of fading channels with channel side information," IEEE Trans. Inform. Theory, vol. 43, pp. 1986-1992, Nov. 1997.

[3]   A. J. Goldsmith and S.-G. Chua, "Variable-rate variable-power MQAM for fading channels," IEEE Trans. Commun., vol. 45, pp. 1218-1230, Oct. 1997.

[4]   M.S. Alouni, X. Tang, and A.J. Goldsmith, "An adaptive modulation scheme for simultaneous voice and data transmission over fading channels," *IEEE J. Sel. Areas Commun.*, vol. 17, no. 5, pp.837-850, May 1999.

[5]   C.S. Hwang and Y. Kim, "An adaptive modulation for integrated voice/data traffic over Nakagami fading channels," *IEEE VTC Fall*, Atlantic City, NJ, Sep. 2001, pp. 1649-1652.

[6]   M. Pursley and J. Shea, "Nonuniform Phase-Shift-Key Modulation for Multimedia Multicast Transmission in Mobile Wireless Networks," IEEE Journal on Selected Areas   in Communications, vol. 17, pp. 774-783, May 1999.

[7]   M. Hossain, P. Vitthaladevuni, M. Alouni, V. Bhargava and, A. Goldsmith, "Adaptive hierarchical Modulation for Simultaneous Voice and Multiclass Data Transmission over Fading Channels," IEEE Transactions on Vehicular Technology, vol. 55, March 2006.

[8]   T. Cover, "Broadcast Channels," *IEEE Transactions on Information Theory*, vol. IT-18, January 1972, pp. 2-14.

[9]   C. W. Sundberg, W. C. Wong, and R. Steele, "Logarithmic PCM Weighted QAM Transmission over Gaussian and Rayleigh Fading Channels," *IEE Proceedings*, Pt. F, vol. 134, pp. 557-570, October 1987.

[10]   R. Cox, J. Hagenauer, N. Seshadri and C. W. Sundberg, "Sub-band Speech Coding and Matched Convolution Channel Coding for Mobile Radio Channels," *IEEE Transactions   on   Signal Processing*, vol. 39, Aug. 1991, pp. 1717-1731.

[11]   Jing Lu, A. Annamalai and D. R. Vaman, "Asymmetric PSK   Constellation Design to Minimize Distortion in PCM Data Transmission," *Proc. IEEE Sarnoff Symposium*, April 2008.

[12]   K. Ramachandran, A. Ortega, K. Uz and M. Vetterli, "Multiresolution Broadcast for Digital HDTV using Joint Source/Channel Coding," *IEEE Journal on Selected Areas in Communications*, vol. 11, Jan. 1993, pp. 6-23.

[13]   ETSI EN-300-744, Digital Video Broadcasting: Framing Structure, Channel Coding, and Channel Coding for Digital Terrestrial Television, 2001.

[14]   A. Calderbank and N. Seshadri, "Multilevel Codes for Unequal Error Protection," *IEEE Transactions on Information Theory*, vol. 39, July 1993, pp. 1234-1248.

[15]   Jing Lu, A. Annamalai, D. R. Vaman, "Reducing Signal Distortion Due to Transmission Errors via Multiresolution Digital Modulations," *Proc. 2008 Wireless Telecommunications Symposium*, California, April 2008.

[16]   M. Sajadieh, F. Kschischang, and A. Leon-Garcia, "Modulation Assisted Unequal Error Protection over Fading Channel," *IEEE Trans. on Vehicular Technology*, vol. 47, Aug. 1998, pp. 900-908.

[17]   M. Pursley and J. Shea, "Multimedia Multicast Wireless Communications with Phase-Shift-Key Modulation and Convolutional Coding," *IEEE Journal on Selected Areas in Communications*, vol. 17, Nov. 1999, pp. 1999-2010.

[18]   M. Pursley and J. Shea, "Adaptive Nonuniform Phase Shift   Key Modulation for Multimedia Traffic in Wireless Networks," *IEEE Journal on Selected Areas in Communications*, vol. 18, August 2000, pp. 1394-1407.

[19]   P. Vitthaladevuni and M. Alouini, "Exact BER Computation of Generalized Hierarchical PSK Constellations," *IEEE Transactions on Communications*, vol. 51, December 2003, pp. 2030-2037.

[20]   P. Vittahaladevuni and M. S. Alouini, "A recursive algorithm for the exact BER computation of generalized hierarchical QAM constellations," *IEEE Trans. Info. Theory*, Jan. 2003, pp. 297-307.

[21]   J. Lassing, E. Strom, E. Agrell and T. Ottosson, "Computation of Exact Bit Error Rate of Coherent M-ary PSK with Gray Code Bit Mapping," *IEEE Transactions on Communications*, 2003.

[22]   J. Liu and A. Annamalai, "Multiresolution Signaling for Multimedia Multicasting," *Proc. 2004 IEEE Vehicular Technology Conference*, Fall 2004, pp. 1088-1092.

[23]   A. Annamalai and J. Liu, "A Cross-Layer Design Perspective for Multiresolution Signaling," *Proc. 2004 IEEE GLOBECOM*, November 2004, pp. 3342-3346.

[24]   A. Annamalai, Jing Lu, D. Vaman, "A Receiver-Oriented Rate-Adaptation Strategy for Improving Network Efficiency in Mobile Ad-Hoc Networks," *Proc. 2008 IEEE Sarnoff Symposium*, Princeton, April 2008.

[25]   A. Annamalai, Jing Lu, D. Vaman, "Improving the Efficiency of Wireless Networks via a Passive Rate-Adaptation Strategy," *Proc. Wireless Telecommunications Symposium*, California, April 2008.

APPENDIX

A summary of coefficients for average and individual bit PER probability using exponential approximation for 4-PSK, 8-PSK, 16-PSK, 64-QAM, and 256-QAM are listed in the following tables.

Table A-1. POLYNOMIAL COEFFICIENTS FOR 4-PSK (INDIVIDUAL BIT CLASS) PROBABILITY

		P_4	P_3	P_2	P_1	P_0
1	$a_\beta^{(1)}$	N/a	-52.3318	44.2865	-11.2667	131.2686
	$b_\beta^{(1)}$	N/a	2.1061	-3.3530	0.1288	1.0276
	$c_\beta^{(1)}$	N/a	1.1291e+04	-9.7238e+03	2.5352e+03	-4.871e+03
	$d_\beta^{(1)}$	-93.3600	114.7000	-35.7200	7.6130	5.5500
2	$a_\beta^{(2)}$	N/a	1.0698e+03	-1.3137e+03	529.1871	60.9991
	$b_\beta^{(2)}$	N/a	-2.1140	3.3542	-0.1252	0.0061
	$c_\beta^{(2)}$	N/a	-1.2594e+05	1.5337e+05	-6.1626e+04	3.4314e+03
	$d_\beta^{(2)}$	415.4000	-643.7000	405.6000	-140.8000	32.5700

Table A-2. POLYNOMIAL COEFFICIENTS FOR 4-PSK (AVERAGE BIT CLASS) PROBABILITY

	P_4	P_3	P_2	P_1	P_0
a_β	N/a	0	0	0	127.4544
b_β	N/a	0	0	0	0.5161
c_β	N/a	0	0	0	-4.4866e+03
d_β	639.0000	-868.8000	487.7000	-153.5000	32.5100

Table A-3. POLYNOMIAL COEFFICIENTS FOR 8-PSK (INDIVIDUAL BIT CLASS) PROBABILITY

		P_4	P_3	P_2	P_1	P_0
1	$a_\beta^{(1)}$	N/a	3.3107e+03	-2.6421e+03	416.5972	104.9653
	$b_\beta^{(1)}$	N/a	11.3557	-13.4411	2.4242	0.8750
	$c_\beta^{(1)}$	N/a	-1.0933e+05	7.9648e+04	-7.3081e+03	-3.6766e+03
	$d_\beta^{(1)}$	270.8000	-155.6000	46.3600	-3.3020	6.0670

2	$a_\beta^{(2)}$	N/a	-495.0799	515.9588	-176.1712	81.9131
	$b_\beta^{(2)}$	N/a	-1.7758	1.0259	0.2603	-0.0148
	$c_\beta^{(2)}$	N/a	1.7697e+04	-1.8307e+04	6.2392e+03	-1.6219e+03
	$d_\beta^{(2)}$	-900.5000	1013.0000	-303.8000	-16.7000	27.0100
3	$a_\beta^{(3)}$	N/a	1.3104e+03	-2.2445e+03	1.1351e+03	-99.8683
	$b_\beta^{(3)}$	N/a	2.2520	-0.7339	0.0943	-0.0042
	$c_\beta^{(3)}$	N/a	-1.3551e+04	2.8639e+04	-1.6005e+04	1.4410e+03
	$d_\beta^{(3)}$	563.8000	-947.4000	648.7000	-251.3000	61.0100

Table A-4. POLYNOMIAL COEFFICIENTS FOR 8-PSK (AVERAGE BIT CLASS) PROBABILITY

	P_4	P_3	P_2	P_1	P_0
a_β	N/a	0	0	0	93.6387
b_β	N/a	0	0	0	0.1529
c_β	N/a	0	0	0	-2.4908e+03
d_β	6.3440e-11	166.7000	-50.0000	-76.67	44.75

Table A-5. POLYNOMIAL COEFFICIENTS FOR 64-QAM (INDIVIDUAL BIT CLASS) PROBABILITY

		P_4	P_3	P_2	P_1	P_0
1	$a_\beta^{(1)}$	N/a	-910.0549	1.1941e+03	-514.9076	107.6440
	$b_\beta^{(1)}$	N/a	2.0915	-0.8068	-1.1349	0.5335
	$c_\beta^{(1)}$	N/a	3.6149e+04	-4.2578e+04	1.6469e+04	-2.4243e+03
	$d_\beta^{(1)}$	215.9000	-124.9000	54.8100	0.05396	8.8710
2	$a_\beta^{(2)}$	N/a	-938.0158	845.9982	-221.7132	78.8082
	$b_\beta^{(2)}$	N/a	-0.3735	0.1735	0.0628	-0.0038
	$c_\beta^{(2)}$	N/a	5.2979e+04	-4.6210e+04	1.1348e+04	-1.6320e+03
	$d_\beta^{(2)}$	559.2000	-818.3000	491.3000	-147.7000	39.8500
3	$a_\beta^{(3)}$	N/a	5.5567e+03	-5.5940e+03	1.7888e+03	-63.3358
	$b_\beta^{(3)}$	N/a	0.3320	-0.0860	0.0100	-4.3182e-04
	$c_\beta^{(3)}$	N/a	-4.2313e+05	4.1430e+05	-1.2817e+05	9.4184e+03
	$d_\beta^{(3)}$	785.4000	-1265.0000	816.1000	-286.7000	70.6500

Table A-6. POLYNOMIAL COEFFICIENTS FOR 64-QAM (AVERAGE BIT CLASS) PROBABILITY

	P_4	P_3	P_2	P_1	P_0
a_β	N/a	0	0	0	80.1406
b_β	N/a	0	0	0	0.0247
c_β	N/a	0	0	0	-1.8278e+03
d_β	1181.0000	-1671.0000	958.9000	-306.3000	70.2100

Table A-7. POLYNOMIAL COEFFICIENTS FOR 16-PSK (INDIVIDUAL BIT CLASS) PROBABILITY

		P_4	P_3	P_2	P_1	P_0
1	$a_\beta^{(1)}$	N/a	3.0777e+03	-3.1923e+03	886.7117	-4.9225
	$b_\beta^{(1)}$	N/a	16.8240	-19.0129	3.8032	0.7708
	$c_\beta^{(1)}$	N/a	-7.9575e+04	7.7277e+04	-1.9794e+04	383.0110
	$d_\beta^{(1)}$	1765.0000	-1630.0000	565.8000	-79.8000	9.9600

	$a_\beta^{(2)}$	N/a	-450.6153	776.0497	-415.7862	103.6366
2	$b_\beta^{(2)}$	N/a	-1.4104	-0.1113	0.5506	-0.0343
	$c_\beta^{(2)}$	N/a	2.4628e+04	-3.1731e+04	1.3658e+04	-2.2255e+03
	$d_\beta^{(2)}$	826.4000	-1027.0000	570.1000	-162.7000	33.9600
	$a_\beta^{(3)}$	N/a	169.9268	-210.3289	108.2366	30.9133
3	$b_\beta^{(3)}$	N/a	-0.1718	0.3988	-0.0907	0.0059
	$c_\beta^{(3)}$	N/a	-5.7246e+04	7.7019e+03	-3.9217e+03	58.3874
	$d_\beta^{(3)}$	879.5000	-1376.0000	867.4000	-287.9000	63.6500
	$a_\beta^{(4)}$	N/a	125.8574	-1.3741e+03	988.4554	-90.4477
4	$b_\beta^{(4)}$	N/a	1.4002	-0.8544	0.1702	-0.0106
	$c_\beta^{(4)}$	N/a	2.1002e+05	-1.5881e+05	2.8226e+04	-1.5010e+03
	$d_\beta^{(4)}$	1426.0000	-2171.0000	1317.0000	-443.2000	95.0200

Table A-8. POLYNOMIAL COEFFICIENTS FOR 16-PSK (AVERAGE BIT CLASS)
PROBABILITY

	P_4	P_3	P_2	P_1	P_0
a_β	N/a	0	0	0	44.7276
b_β	N/a	0	0	0	0.0379
c_β	N/a	0	0	0	5.7858
d_β	1892.0000	-2684.0000	1521.0000	-477.8000	95.4300

Table A-9. POLYNOMIAL COEFFICIENTS FOR 256-QAM (INDIVIDUAL BIT CLASS)
PROBABILITY

		P_4	P_3	P_2	P_1	P_0
	$a_\beta^{(1)}$	N/a	214.1059	199.7885	-303.1980	94.6347
1	$b_\beta^{(1)}$	N/a	4.1234	-2.5915	-0.7370	0.5077
	$c_\beta^{(1)}$	N/a	2.1984e+04	-3.0988e+04	1.4448e+04	-2.3255e+03
	$d_\beta^{(1)}$	1664.0000	-1553.0000	558.3000	-73.9800	12.6500
	$a_\beta^{(2)}$	N/a	77.8998	224.0731	-233.3701	83.0034
2	$b_\beta^{(2)}$	N/a	-0.1363	-0.1557	0.1407	-0.0089
	$c_\beta^{(2)}$	N/a	-2.3975e+04	1.7334e+04	-2.0925e+03	-703.7629
	$d_\beta^{(2)}$	619.0000	-753.4000	450.0000	-139.6000	39.3500
	$a_\beta^{(3)}$	N/a	939.0930	-1.3508e+03	572.5007	-8.7737
3	$b_\beta^{(3)}$	N/a	-0.0857	0.1073	-0.0229	0.0014
	$c_\beta^{(3)}$	N/a	2.5014e+04	72.5132	-8.7253e+03	766.6995
	$d_\beta^{(3)}$	955.7000	-1446.0000	890.6000	-288.9000	70.4800
	$a_\beta^{(4)}$	N/a	-1.8191e+03	385.9918	663.9760	-85.3526
4	$b_\beta^{(4)}$	N/a	0.2151	-0.1277	0.0251	-0.0016
	$c_\beta^{(4)}$	N/a	2.0229e+05	-1.8229e+05	3.6211e+04	-1.8963e+03
	$d_\beta^{(4)}$	1494.0000	-2260.0000	1362.0000	-450.6000	102.4000

Table A-10. P OLYNOMIAL COEFFICIENTS FOR 256-QAM (AVERAGE BIT CLASS)
PROBABILITY

	P_4	P_3	P_2	P_1	P_0
a_β	N/a	0	0	0	63.2383
b_β	N/a	0	0	0	0.0061
c_β	N/a	0	0	0	-1.1143e+03
d_β	1735.0000	-2475.0000	1426.0000	-458.30000	101.0000

4

MULTIUSER BER ANALYSIS OF CS-QCSK MODULATION SCHEME IN A CELLULAR SYSTEM

K.Thilagam[1] and K.Jayanthi[2]

[1] Research Scholar, Department of ECE, Pondicherry Engg. College, Puducherry, India
thilagam.k@pec.edu

[2]Associate Professor, Department of ECE, Pondicherry Engg. College, Puducherry,India
jayanthi@pec.edu

ABSTRACT

In recent years, chaotic communication is a hot research topic and it suits better for the emerging wireless networks because of its excellent features. Different chaos based modulation schemes have evolved, of which the CS-DCSK modulation technique provides better BER performance and bandwidth efficiency, due to its code domain approach. The QCSK modulation technique provides double benefit: higher data rate with similar BER performance and same bandwidth occupation as DCSK. By combining the advantage of code shifted differential chaos shift keying (CS-DCSK) and Quadrature chaos shift keying (QCSK) scheme, a novel modulation scheme called code shifted Quadrature chaos shift keying (CS-QCSK) is proposed and its suitability in a multiuser scenario is tested in this paper. The analytical expressions for the bit-error rate for Multi-user CS-QCSK scheme (MU-CS-QCSK) under Rayleigh multipath fading channel is derived. The simulation result shows that, in multiuser scenario the proposed method outperforms classical chaotic modulation schemes in terms of bit error rate (BER).

KEYWORDS

Differential chaos shift keying (DCSK) ; Quadrature chaos shift keying (QCSK); Code shifted differential chaos shift keying (CS-DCSK); code shifted Quadrature chaos shift keying (CS-QCSK); Multiuser code shifted Quadrature chaos shift keying (MU-CS-QCSK); Bit Error Rate(BER).

1. INTRODUCTION

In the last few years, a great research effort has been devoted, concerning chaotic carrier modulation. In contrary to the conventional modulation schemes, a wideband, non-periodic chaotic signal is used as carrier, which results in good correlation characteristics and robustness against multipath fading effects. The basic concept of chaos based modulation techniques for coherent and non-coherent systems have been analyzed and studied in [1] &[2]. For a coherent system, the receiver needs an exact replica of the chaotic signal, such robust synchronization techniques are still not realizable in a practical environment [3].It is worth mentioning the evolution of various types chaos based modulation schemes and their significance and this section is devoted for summarizing the literature survey/ research findings of such schemes. The DCSK scheme, discussed in [4] represents a more robust non-coherent scheme, in which the exact knowledge of the chaotic basis functions is not needed at the receiver. To further enhance the DCSK scheme, frequency-modulated DCSK (FM-DCSK) was introduced to overcome the varying bit-energy problem in DCSK, which is very well explored in [5], [6]. The limitation of the work is that the analysis was purely simulation based with a constant gain channel model.

In [7], the role of synchronization in digital communications using chaos was discussed. At receiver, the two pieces of chaotic signals are correlated, and then the binary symbol is decoded based on the sign of the correlator output. The DCSK communication schemes, which exploit the repetitive nature of the chaotic signal, extending from simple binary communications to more sophisticated detectors, have been discussed [8]. In [9], a discrete-time approach for CSK system with multiple users and the multiple-access DCSK (MA-DCSK) system was investigated in depth and a mixed analysis-simulation (MAS) technique was developed to calculate the bit error rates (BERs) using the derived BER analytical expressions. Multipath performance of binary DCSK systems using a standard correlation detector was discussed in [10]-[11]. Quadrature CSK (QCSK) is a multilevel version of DCSK, based on the generation of an orthogonal of chaotic functions. It allows an increase in data rate occupying same bandwidth with respect to DCSK [12]. In [13]-[15], the performance of the DCSK system over a channel with Rayleigh fading and Ricean fading was discussed with the necessity to model the effects of multipath delay spread as well as fading. A multiple-access technique with differential chaos shift keying using a one dimensional iterative map to generate the chaotic signals for all users and similar average data rates for the users have been discussed in [16]-[17]. In [18], the generation of inherent wideband signal with constant energy per bit, which can be achieved by using the combination of frequency modulation with QCSK was discussed. The high-level constellation in FM-QACSK which can improve the speed of chaos shift keying is examined in [19]. In [20], the performance analysis and optimization of multiuser differential chaos shift keying, using dynamically improved gaussian approximation(DIGA) method was discussed.

The DCSK system incorporated with two-user cooperative diversity technique and multiple accessing, using the features of Walsh code sequences was discussed in[21]. It was discussed with the signal structure of chaotic pulse cluster resolving the technical problem and offering a desired solution to multi-user's communication systems using Walsh function division scheme [22]. In[23]a high-efficiency differential chaos shift keying (HEDCSK) with new correlation-based communication scheme, was introduced as an enhanced version of DCSK allowing an increased data rate compared to DCSK with the same bandwidth occupation, but has poor BER performance and bandwidth efficiency. CS-DCSK uses code domain approach, the reference and the information bearing signals are transmitted at the same time slots, it offers bandwidth efficiency but poor BER performance, as discussed in [24]. In [25], the authors discussed that, using the generalized code-shifted DCSK (GCS-DCSK) modulation scheme, the CS-DCSK was extended to transmit multiple bit streams by means of one reference wavelet but has poor BER performance. In [26], a novel chaos based modulation scheme CS-QCSK was proposed and was thoroughly analysed for single user scenario.

In this paper, the analysis of multiuser performance of a novel CS-QCSK modulation scheme is proposed, which is the blend of CS-DCSK and QCSK modulation schemes. In addition to high data rates, it offers the advantage of both the CS-DCSK and QCSK schemes [24], [12]. Analytical expressions for computing BER for the proposed approach has been derived. Further simulation results prove its supremacy over the classical chaos modulation schemes.

The remaining part of the paper is organized as follows: In section 2, conventional schemes and the proposed system model with problem formulation is described. Section 3, elaborates the analytical BER derivation for Rayleigh fading channel. Simulation results are discussed in Section 4. Section 5, deals with the conclusion of the paper.

2. OVERVIEW ON CHAOS BASED MODULATION SCHEMES

In order to explain the MU-CS-QCSK system, a brief introduction to DCSK, QCSK, CS-DCSK and CS-QCSK methods is essential and details are given in this section.

2.1. Differential Chaos Shift Keying (DCSK)

In DCSK system, for each symbol period, two chaotic sample functions are used to represent, one bit of information [4]. The transmitted DCSK signal has two parts; the first part of the signal is the reference signal, while the next part denotes the information-bearing signal. If the bit '1' is to be transmitted, then the information bearing signal is the replica of the reference part. If the bit '0' is to be transmitted, then the information bearing signal is the inverted copy of the reference part. The DCSK transmitted symbol can be represented as,

$$S_{DCSK}(t) = \begin{cases} x(t) & 0 \le t < \dfrac{T}{2} \\ \mp x(t - T/2) & \dfrac{T}{2} \le t < T \end{cases} \tag{1}$$

where, 'T' is the symbol duration. In the receiver, the original information is extracted by computing the correlation between two received sample functions. The output of the correlator is sampled over each symbol period and the output is finally given to the decision device to determine the binary values 1's or 0's.

2.2. Quadrature Chaos Shift Keying (QCSK)

Similar to DCSK method, each symbol period of QCSK also has two parts, but the modification is, information-bearing signals holds two bits of information by means of quadrature chaos shift keying technique. Since two bits of information are transmitted, it is possible for QCSK scheme to obtain higher data rate [12]. Quadrature chaotic signals are produced by an orthonormal basis chaotic sample functions x(t) and y(t). Let, x(t) be the chaotic reference signal , assuming that the signal x has zero mean value and is defined as

$$x(t) = \sum_{k=1}^{\infty} f_k \sin(k \omega t + \varphi_k) \tag{2}$$

and let y(t) be the quadrature chaotic reference signal, obtained by changing the phase of each fourier frequency component by $\pi/2$, and is defined as,

$$y(t) = \sum_{k=1}^{\infty} f_k \sin(k \omega t + \varphi_k - \pi/2) \tag{3}$$

The signals x(t) and y(t) are orthogonal in the interval $I_t = [0,t]$ and is defined as,

$$x \perp y \Leftrightarrow \int_0^\tau x(t) y(t) dt = 0 \tag{4}$$

The transmitted QCSK signal is given by,

$$S_{QCSK}(t) = \begin{cases} \sqrt{E_b} c_x(t) & 0 < t < \dfrac{T}{2} \\ \sqrt{E_b}(a_1 c_x(t - T/2) + a_2 c_y(t - T/2)) & \dfrac{T}{2} \le t < T \end{cases} \tag{5}$$

where E_b is the bit energy over $T/2$ period, satisfying the orthogonal relations.

$C_x(t-T/2)$ is chaotic signal with duration T/2 and $C_y(t-T/2)$ is chaotic signal orthogonal to $C_x(t-T/2)$ with same duration of T/2.

2.3. Code Shifted - Differential Chaos Shift Keying (CS-DCSK)

In DCSK, the reference and the information bearing signals are transmitted in two consecutive time slots because of its TDMA approach. This time domain approach requires two independent channels for the transmission of reference and information bearing signals. Further, it requires a delay component both in the modulator and demodulator circuits and halves the data rate. To overcome this drawback, an alternative approach used is CS-DCSK, where both the reference and the information bearing signals are transmitted in the same time slot because of its code domain approach (i.e) the two signals are separated by walsh codes instead of time delay [25]. The transmitted CS-DCSK signal is given by,

$$S_{CS-DCSK}(t) = \sum_{k=0}^{N-1} W_{R,k+1} C(t - kT_c) + a \sum_{k=0}^{N-1} W_{I,k+1} C(t - kT_c) \tag{6}$$

Where, $T_s = NT_c$, $W_{R,k+1}$ is the reference signal, $W_{I,k+1}$ is the information signal, $C(t-kT_c)$ is the chaotic signal, T_s is the symbol duration and T_c is the chip duration. The limitation of this system is, with the increased complexity the bit error rate obtained by this system is more or less similar when compared with the existing system.

2.4. Code Shifted - Quadrature Chaos Shift Keying (CS-QCSK)

The aforementioned CS-DCSK has poor BER performance and minimum data rate. To overcome this drawback, a novel chaos based modulation scheme Code shifted-Quadrature Chaos Shift Keying (CS-QCSK) is proposed. In this scheme, both the reference and the information bearing signals are transmitted in the same time slot, using walsh codes and the difference is information bearing part holds two bits of information. CS-QCSK can also be said as Hybrid of QCSK and CS-DCSK which will hold the advantage of both the schemes, so that it provides a higher data rate, bandwidth efficiency and better BER performance. The CS-QCSK transmitted signal is given by,

$$S_{CS-QCSK}(t) = \sum_{k=0}^{N-1} W_{R,k+1} C_x(t - kT_c) + a_1 \sum_{k=0}^{N-1} W_{I1,k+1} C_x(t - kT_c) + a_2 \sum_{k=0}^{N-1} W_{I2,k+1} C_y(t - kT_c) \tag{7}$$

Where, $T_s = NT_c$, $W_{I1,k+1}$ and $W_{I2,k+1}$ are the information signal, $C_x(t-kT_c)$ is the chaotic signal and the $C_y(t-kT_c)$ is the orthogonal chaotic signal, T_s is the symbol duration and T_c is the chip duration. The limitation of this system is its increased complexity and the performance of the system begins to deteriorate for large number of users. The next section briefly explains the multiuser CS-QCSK system model and its performance analysis in the Rayleigh multipath environment.

2.5. System Model of Multiuser CS-QCSK Modulation

The aim of this section is to illustrate the practical importance of CS-QCSK modulation scheme, whose simplified transmitter and receiver block diagram are shown in figures 1 and 2 respectively. A baseband system is considered for simplicity. But it is understood, that if the scheme is to be employed for wireless communications a modulator is in need to generate the corresponding RF passband signal. Furthermore, it is assumed that the description of the MU-CS-QCSK scheme in the continuous-time domain admits an equivalent discrete-time representation. From the earlier stated discussions, it can be concluded that, by combining CS-DCSK and QCSK schemes, a novel CS-QCSK modulation scheme can be realised and it is thoroughly analysed in the presence of multiple users.

2.5.1. Transmitter

In MU-CS-QCSK system, the symbol 'S' is transmitted with the reference signal and information signal in the same time slot but separated by Walsh code sequences. Let there be 'N_u' number of users.

The MU-CS-QCSK transmitted signal is given by,

$$S_M(t) = \sum_{k=0}^{N-1} W_{R,k+1} C_x(t - kT_c) + \sum_{u=1}^{N_u} \left[a_{1u} \sum_{k=0}^{N-1} W_{I1u,k+1} C_x(t - kT_c) + a_{2u} \sum_{k=0}^{N-1} W_{I2u,k+1} C_y(t - kT_c) \right] \tag{8}$$

Where, $T_s=NT_c$, $a_{1u}\in\{-1, 1\}$, $a_{2u}\in\{-1, 1\}$ is mapped from $b\in\{0, 1\}$ which is the information bit to be transmitted. This scheme uses different Walsh code for the reference and information signal, where $W_{R,k+1}$ represent the Walsh code for reference signal and $W_{I1u,k+1}$, $W_{I2u,k+1}$ represent the Walsh code for the two consecutive information bits, C(t) is the chaotic signal with duration of T_c and u=1,....,N_u represents the number of users in the system. Both the reference and the information signal are transmitted in the same time slot as given in equation (8). The orthogonality of two signals are given by,

$$C_x\left(t-kT_c\right)=\begin{cases}c_x\left(t\right) & kT_c\le t<(k+1)T_c \\ 0 & otherwise\end{cases} \tag{8.a}$$

$$C_y\left(t-kT_c\right)=\begin{cases}c_y\left(t\right) & kT_c\le t<(k+1)T_c \\ 0 & otherwise\end{cases} \tag{8.b}$$

Let,

$$\Delta=\int_0^{T_s=NT_c}\sum_{k=0}^{N-1}W_{R,k+1}C_x\left(t-kT_c\right)a_{1u}\sum_{k=0}^{N-1}W_{I1u,k+1}C_x\left(t-kT_c\right)a_{2u}\sum_{k=0}^{N-1}W_{I2u,k+1}C_y\left(t-kT_c\right).dt \tag{9}$$

$$E_b=\int_0^{T_s}c^2(t)c_y(t).dt \tag{9.a}$$

$$(.)^T=Transporse\ of\ vector.$$

$$c_x\left(t\right)=C_x\left(t-kT_c\right)\ ;\ \forall k \tag{9.b}$$

$$c_y\left(t\right)=C_y\left(t-kT_c\right)\ ;\ \forall k \tag{9.c}$$

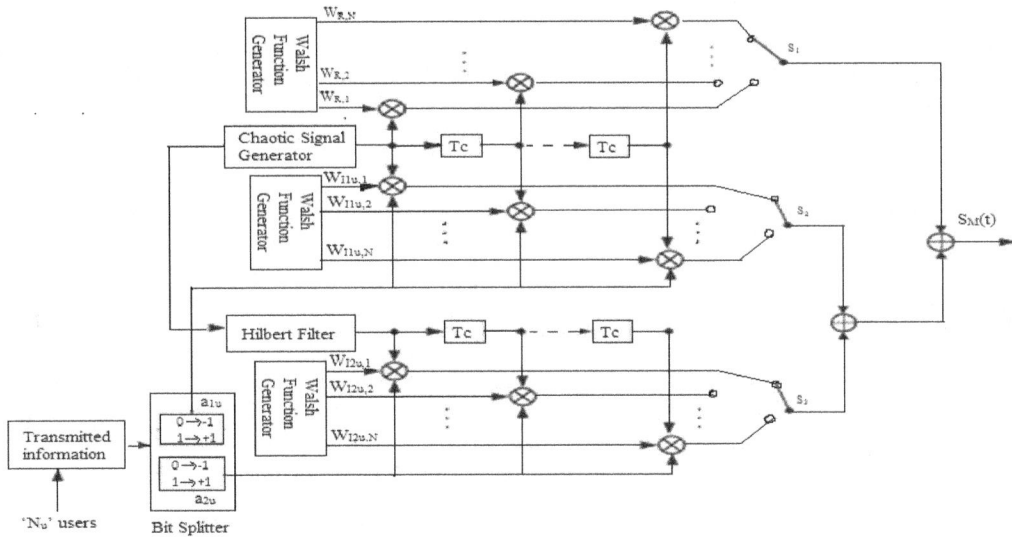

Figure 1. Block diagram of Multiuser CS-QCSK Transmitter

Referring to the transmitter block diagram of figure 1, the CS-QCSK modulation scheme consist of chaos signal generator, which generates chaos signal C_x (t) and its orthogonal signal C_y (t).The orthogonal signal is obtained by taking the Hilbert transform of the chaos signal C_x (t). The transmitter block consists of three Walsh function generator, which generates three different Walsh code sequences that are orthogonal in nature. It has (N-1) delay elements of T_c

chip duration. The three Walsh function generator, generate Walsh codes, $W_{R,k+1}$ for reference signal, and $W_{I1u,k+1}$, $W_{I2u,k+1}$ for information signal for 'u' users, where, u=1,....Nu. The reference signal of Walsh code $W_{R,k+1}$ is multiplied with the chaos signal and the corresponding output is obtained by the switching unit S1.It has a bit splitter, which splits the input bit stream into odd and even sequences. The first information signal of the uth user's Walsh code '$W_{I1u,k+1}$' is multiplied with the odd sequence of bits and the chaos signal ,the corresponding output is obtained by the switching unit S_2. The second information signal of the uth user's walsh code '$W_{I2u,k+1}$' is multiplied with the even sequence of bits and the orthogonal chaos signal, the corresponding output is obtained by the switching unit S_3.The output obtained at the switching unit S_2 and S_3 are summed for 'u' users and it is further summed with the switching unit S1 to obtain the final output. The composite CS-QCSK signal is finally obtained.

2.5.2. Receiver

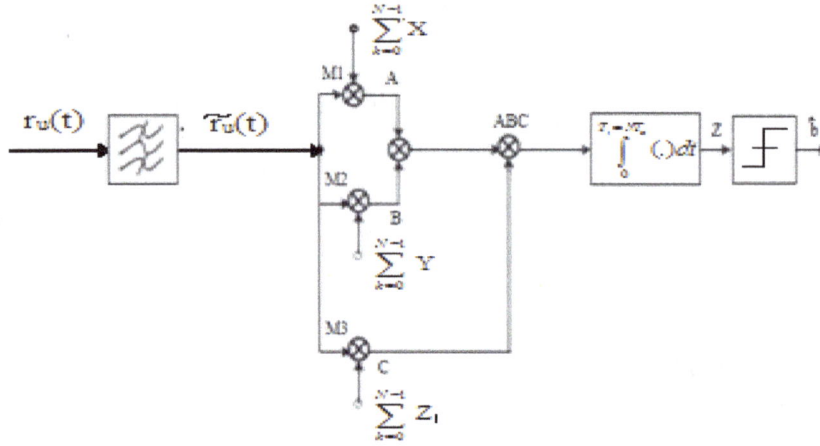

Figure 2. Block diagram of MU-CS-QCSK Receiver

Figure 2. shows the possible hardware realisation of MU-CS-QCSK receiver. The receiver unit consists of band pass filter with the bandwidth of 2B which can pass the received signal $r_u(t)$ without any distortion. The receiver filter output with Additive White Gaussian Noise (AWGN) n(t) is obtained as

$$r_u(t) = \sum_{k=0}^{N-1} W_{R,k+1} C_x (t-kT_c) + \sum_{k=0}^{N_u} \left[a_{1u} \sum_{k=0}^{N-1} W_{I1u,k+1} C_x (t-kT_c) + a_{2u} \sum_{k=0}^{N-1} W_{I2u,k+1} C_y (t-kT_c) \right] + n(t) \quad (10)$$

$$\text{rect}_x (t-kT_c) = \begin{cases} 1 & , kT_c \le t < (k+1)T_c \\ 0 & , \text{otherwise} \end{cases} \quad (10.a)$$

$$\text{rect}_y (t-kT_c) = \begin{cases} 1 & , kT_c \le t < (k+1)T_c \\ 0 & , \text{otherwise} \end{cases} \quad (10.b)$$

Let,

$$X = W_{R,k+1} \text{rect}_x (t-kT_c) , \quad Y = W_{I1u,k+1} \text{rect}_x (t-kT_c) , \quad Z_1 = W_{I2u,k+1} \text{rect}_y (t-kT_c) \quad \text{and}$$

Let,

$$A = S_M(t) \sum_{k=0}^{N-1} W_{R,k+1} \text{rect}_x (t-kT_c) \quad (11)$$

$$B = S_M(t) \sum_{k=0}^{N-1} W_{I1u,k+1} \text{rect}_x (t - kT_c) \tag{12}$$

$$C = S_M(t) \sum_{k=0}^{N-1} W_{I2u,k+1} \text{rect}_y (t - kT_c) \tag{13}$$

The integrator output is thus,

$$Z = \int_0^{T_s = NT_c} ABC.dt \tag{14}$$

Substituting, equation (8) in (11),(12)and(13). The equations (15), (16) and (17) are obtained.

$$A = \sum_{k=0}^{N-1} C_x (t - kT_c) + a_{1u} \sum_{k=0}^{N-1} W_{R,k+1} W_{I1u,k+1} C_x (t - kT_c) + a_{2u} \sum_{k=0}^{N-1} W_{R,k+1} W_{I2u,k+1} C_y (t - kT_c) \tag{15}$$

$$B = \sum_{k=0}^{N-1} W_{R,k+1} W_{I1u,k+1} C_x (t - kT_c) + a_{1u} \sum_{k=0}^{N-1} C_x (t - kT_c) \tag{16}$$

$$C = \sum_{k=0}^{N-1} W_{R,k+1} W_{I2u,k+1} C_x (t - kT_c) + a_{2u} \sum_{k=0}^{N-1} C_y (t - kT_c) \tag{17}$$

Substituting equation (14), (15) and (16) in (13), the observation vector 'Z' is obtained as,

$$Z = \int_0^{T_s = NT_c} \left(\sum_{k=0}^{N-1} C_x (t - kT_c) + a_{1u} \sum_{k=0}^{N-1} W_{R,k+1} W_{I1u,k+1} C_x (t - kT_c) + a_{2u} \sum_{k=0}^{N-1} W_{R,k+1} W_{I2u,k+1} C_y (t - kT_c) \right).$$
$$\left(\sum_{k=0}^{N-1} W_{R,k+1} W_{I1u,k+1} C_x (t - kT_c) + a_{1u} \sum_{k=0}^{N-1} C_x (t - kT_c) \right) \left(\sum_{k=0}^{N-1} W_{R,k+1} W_{I2u,k+1} C_x (t - kT_c) + a_{2u} \sum_{k=0}^{N-1} C_y (t - kT_c) \right) dt \tag{18}$$

The above equation (18) can be simplified as,

$$Z \cong a_{1u} a_{2u} E_b \tag{19}$$

3. BER ANALYSIS FOR MULTIPATH RAYLEIGH FADING SCENARIO

The analytical expression for the BER of MU-CS-QCSK under the multipath Rayleigh fading channel is derived in the following section to validate the simulated results.

3.1. MULTIPATH RAYLEIGH FADING CHANNELS

From the receiver unit, the obtained observation variable 'Z' is given by

$$Z = \sum_{k=0}^{N-1} \int_{kT_c}^{(k+1)T_c} \left[W_{R,k+1} \cdot \tilde{r}_r (t) \right] \left[W_{I1u,k+1} W_{I2u,k+1} \cdot \tilde{r}_r (t) \right] dt \tag{20}$$

$$= \sum_{k=0}^{N-1} W_{R,k+1} W_{I1u,k+1} W_{I2u,k+1} \int_{kT_c}^{(k+1)T_c} \left[\left(W_{R,k+1} + \sum_{u=1}^{N_u} a_{1u} W_{I1u,k+1} + \sum_{u=1}^{N_u} a_{2u} W_{I2u,k+1} \right) . c \left(t - kT_c \right) + n(t) \right]^2 .dt \qquad (21)$$

There are different types of chaotic maps to generate the chaotic signals, they are classified as (i)logistic map (ii) cubic map (iii) Bernoulli-shift map. In the proposed method, logistic map is considered and is defined as, x $(n+1) = 1-2x^2(n)$.Where, $\beta = T_c f_s$ is the number of samples in a chip time and $K = T_s f_s = N\beta$ is the symbol duration. In the remaining part of the paper, 'K' is referred as the spreading factor. Consider, the transmission of $a_{1u} = a_{2u} = +1$ sequence. The logistic map is defined in the continuous time domain. In order to convert into the discrete time domain, consider C_j to be the samples of chaotic signal and η_j be its corresponding channel noise. Then, 'Z' the equation for discrete time equivalent is defined as,

$$Z = \sum_{k=0}^{N-1} W_{R,k+1} W_{I1u,k+1} W_{I2u,k+1} \sum_{m=1}^{\beta} \left[\left(W_{R,k+1} + a_{1u} W_{I1,k+1} + a_{2u} W_{I2u,k+1} \right) . c_{k\beta+m} + \eta_{k\beta+m} \right]^2 \qquad (22)$$

Where,

$$c_{k\beta+m} = \begin{cases} c_m & , \kappa\beta \le \kappa\beta + m < (k+1)\beta \\ 0 & , otherwise \end{cases} \qquad (23)$$

From the equation (22), the decision variable is decomposed into three terms,

$$Z = Z_A + Z_B + Z_C \qquad (24)$$

where,

$$Z_A = \sum_{k=0}^{N-1} W_{R,k+1} W_{I1u,k+1} W_{I2u,k+1} \sum_{m=1}^{\beta} \left[\left(W_{R,k+1} + a_{1u} W_{I1u,k+1} + a_{2u} W_{I2u,k+1} \right)^2 \left(c_{k\beta+m} \right)^2 \right] \qquad (25)$$

$$Z_B = \sum_{k=0}^{N-1} W_{R,k+1} W_{I1u,k+1} W_{I2u,k+1} \sum_{m=1}^{\beta} \left[\left(W_{R,k+1} + a_{1u} W_{I1u,k+1} + a_{2u} W_{I2u,k+1} \right) \left(c_{k\beta+m} \right) \left(\eta_{k\beta+m} \right) \right] \qquad (26)$$

$$Z_C = \sum_{k=0}^{N-1} W_{R,k+1} W_{I1u,k+1} W_{I2u,k+1} \sum_{m=1}^{\beta} \left(c_{k\beta+m} \right)^2 \qquad (27)$$

By applying, expectation and variance to the above equations, the expectation and variance values of these variables are obtained as follows:

$$E\left\{ Z_A \mid b = 1 \right\} = E\left\{ Z_A \mid a = +1 \right\} = 3N\beta + \frac{1}{2} N\beta \left[N_u^2 \left(N_u + 1 \right)^2 \right] E\left\{ C_j^2 \right\} \qquad (28)$$

$$E\left\{ Z_B \mid b = 1 \right\} = E\left\{ Z_B \mid a = +1 \right\} = 0 \qquad (29)$$

$$E\left\{ Z_C \mid b = 1 \right\} = E\left\{ Z_C \mid a = +1 \right\} = 0 \qquad (30)$$

$$Var\left\{ Z_A \mid b = 1 \right\} = Var\left\{ Z_A \mid a = +1 \right\} = 9N^2\beta^2 + \frac{7}{4} N^2\beta^2 N_u^6 Var\left\{ C_j^2 \right\} \qquad (31)$$

$$Var\{Z_B|b=1\}=Var\{Z_B|a=+1\}=4N^2\beta^2N_0^2N_u^2\left(N_u+1\right)^2Var\{C_j^2\} \tag{32}$$

$$Var\{Z_C|b=1\}=Var\{Z_C|a=+1\}=N\beta N_0^2 \tag{33}$$

Where, E{.} represents the expectation operator and var{.} represents the variance operator, then

$$E\{Z|b=1\}=E\{Z|a=+1\}=3N\beta+\frac{1}{2}N\beta\left[N_u^2(N_u+1)^2\right]E\{C_j^2\} \tag{34}$$

$$Var\{Z|b=1\}=Var\{Z|a=+1\}=9N^2\beta^2+\frac{7}{4}N^2\beta^2N_u^6Var\{C_j^2\}+4N^2\beta^2N_0^2N_u^2\left(N_u+1\right)^2Var\{C_j^2\}+N\beta N_0^2 \tag{35}$$

For the transmission of '$a_{1u}=a_{2u}=-1$' sequence, the expectation and variance of observation variable is derived similarly.

$$E\{Z|a=-1\}=-E\{Z|a=+1\} \tag{36}$$

$$Var\{Z|a=-1\}=Var\{Z|a=+1\} \tag{37}$$

It is assumed, that the probability distribution of the observation variable can be approximated and the transmitted bits are equiprobable. Therefore, the bit error rate of MU-CS-QCSK modulation scheme is obtained as,

$$BER=\frac{1}{2}Pr(Z_{mu}<0|b=1)+\frac{1}{2}Pr(Z_{mu}\geq0|b=0) \tag{38}$$

$$=\frac{1}{2}Pr(Z_{mu}<0|a=+1)+\frac{1}{2}Pr(Z_{mu}\geq0|a=-1)$$

$$=\frac{1}{2}erfc\left(\frac{E\{Z_{mu}|a=+1\}}{\sqrt{2Var\{Z_{mu}|a=+1\}}}\right)$$

Where, m=2; represents the number of information bits transmitted per time slot and u=1, ...,Nu; represents the number of users. For logistic maps, the expectation of chaos signal is $E\{C_j^2\}=1/2$ and variance of chaos signal is $Var\{C_j^2\}=1/8$. By substituting, the expectation values in the numerator and variance values in the denominator in the above equation, it becomes

$$=\frac{1}{2}erfc\left(\frac{3N\beta+\frac{1}{2}N\beta\left[N_u^2(N_u+1)^2\right]E\{C_j^2\}}{\sqrt{2\left(9N^2\beta^2+\frac{7}{4}N^2\beta^2N_u^6Var\{C_j^2\}+4N^2\beta^2N_0^2N_u^2\left(N_u+1\right)^2Var\{C_j^2\}+N\beta N_0^2\right)}}\right)$$

$$\tag{39}$$

The conditional BER measured for the multipath channel is given by,

BER (α_0, α_1.......... α_{L-1}) = BER (γ_b) $\tag{40}$

Where, $\gamma_b = (E_b/N_o)$ (α_0, α_1.......... α_{L-1}) = ($\gamma_0 + \gamma_1 +$.................$+ \gamma_{L-1}$).

Let, L be the number of propagation paths, 'α_l' be the gain of the lth path, τ_l be the delay of the lth path. The gains α_l of the propagation paths is assumed to be independent identical distributed Rayleigh random variables. The probability density function (PDF) of γ_i is an exponential distribution given by,

$$f\left(x\right) = \frac{1}{\gamma_k} \exp(\frac{-x}{\gamma_k})$$

(41)

The probability density function (pdf) of γ_b is obtained as,

$$f\left(\gamma_b\right) = f\left(\gamma_0\right) \otimes f\left(\gamma_1\right)\ldots\ldots\ldots\otimes f\left(\gamma_{L-1}\right)$$

(42)

Where, f(γ_k) is the pdf of instantaneous (SNR) measured in the ith path. By averaging the conditional bit error rates in the multipath channels, the total BER measured is obtained as,

$$BER = \int_0^\infty BER\left(\gamma_b\right) f\left(\gamma_b\right) d\gamma_b$$

(43)

It is assumed to have two propagation paths, and then the bit error rate expressions for Multiuser CS-QCSK (MU-CS-QCSK) can be simplified as,

$$BER_{MU-CS-QCSK} = \frac{1}{2} erfc\left(\frac{3K + \frac{1}{2}(\gamma_b) \cdot N_u^2}{\sqrt{\frac{1}{4}\left(9K^2 + \frac{7}{4}K^2 N_u^6\right) + \frac{1}{2} K N_u^2 (\gamma_b) + 2K}}\right)$$

(44)

In order to validate the analytical approach of the proposed scheme, an attempt is made through simulations. The next section presents the simulation analysis of the proposed scheme and it is finally compared with the analytical results.

4. RESULTS AND DISCUSSION

The BER performance of MU-CS-QCSK modulation scheme is analyzed through Matlab simulation. Simulation parameters taken for the analysis are:

Table 1. Simulation Parameters

Parameters	Values / Types
Chip duration(Tc)	2 microsecond
Symbol duration(Ts)	2 millisecond
Transmission power(P)	43dBm
Noise power spectral density(No)	1 dBm
Chaotic Mapping Types	Bernoulli map, Logistic map, Cubic map
Spread factor(K)	4,8,20
Total number of users	100
Fading Channel type	Rayleigh

The BER performance of proposed MU-CS-QCSK scheme for various parameters under multipath Rayleigh fading channel is discussed further with the obtained graphs.

4.1. BER Performance of Multipath Rayleigh Channel

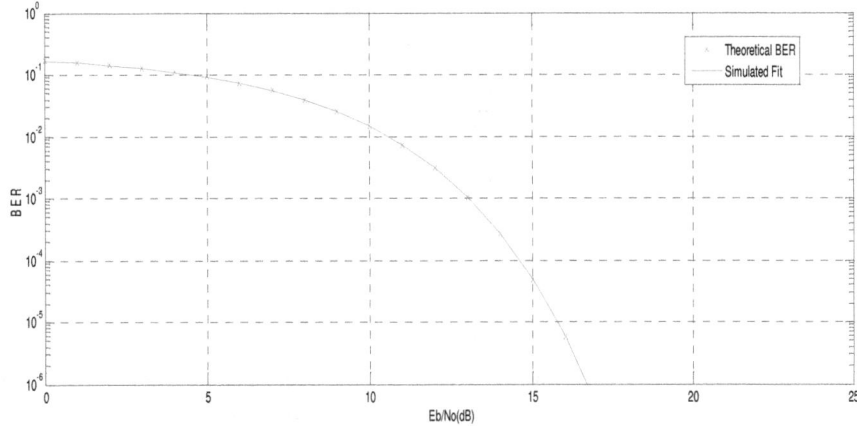

Figure 3. BER performance of MU- CS-QCSK modulation scheme.

From the figure.3, it is inferred that BER performance of MU-CS-QCSK scheme is plotted with the theoretical and simulated values.The BER values calculated from the analytical expression and the simulation are very similar and both the graph merge with each other.For the BER value of 10^{-6}, the required (Eb/No) value is 16.5dB(approximately).

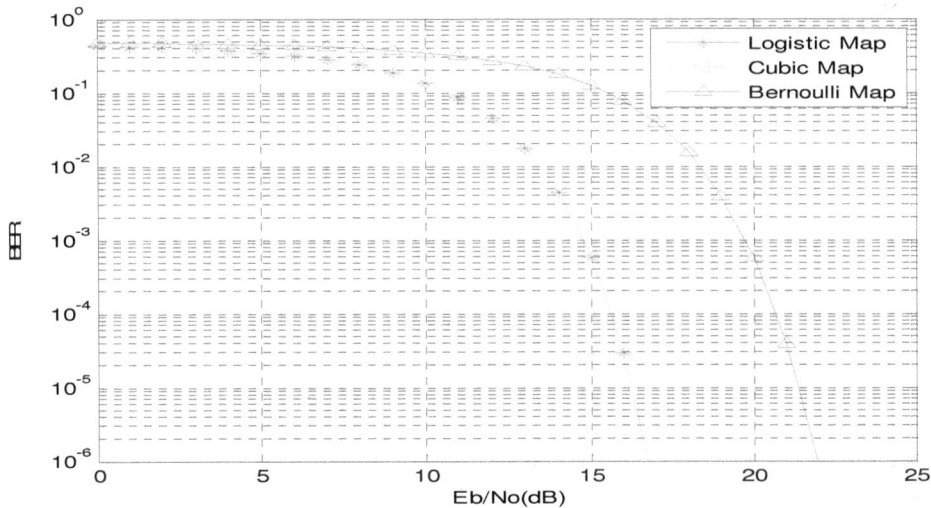

Figure 4. BER performance of MU- CS-QCSK modulation scheme for various chaotic maps

BER performances of MU-CS-QCSK modulation schemes for various chaotic maps were plotted. From the figure, it is inferred that, the logistic map and cubic map has similar and better BER performance compared to the Bernoulli map. For simulation purpose logistic mapping method is used because of its supermacy.

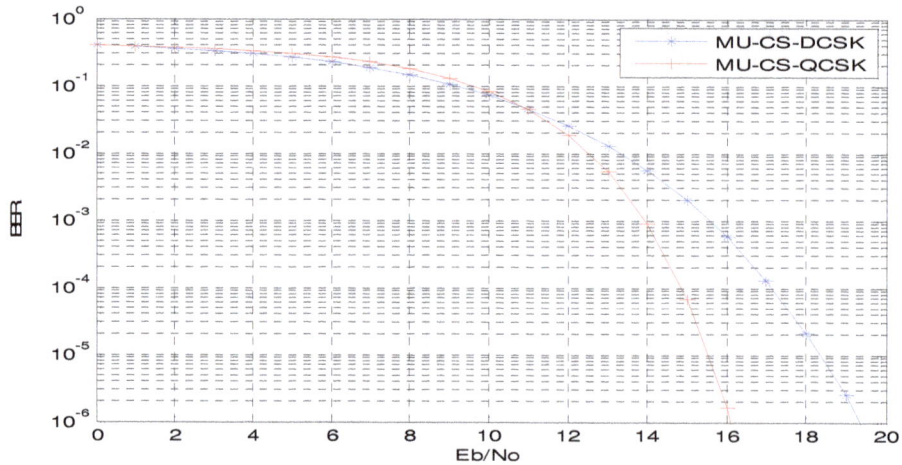

Figure 5. Comparison of BER performance of MU-CS-QCSK modulation scheme

From Figure 5, it can be inferred that, the proposed MU-CS-QCSK scheme offers better BER performance compared to the other multiuser scheme.For the BER of 10-6, the proposed MU-CS-QCSK requires16dB(approx) and theMU-CS-DCSK requires19.1dB(approx).

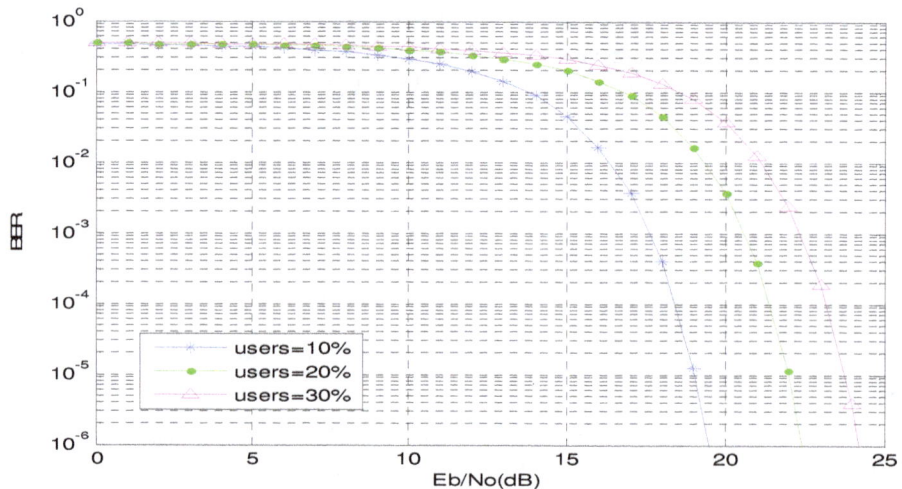

Figure 6. Performance analysis for varying the percentage of users

The BER performance of MU-CS-QCSK modulation scheme for different number of users were plotted. From the graph, it is infered that as the number of user increases the BER performance is limited(i.e) for 10 % ,20% and 30% of the users it requires around 19.1 dB, 22dB and 24dB respectively.This is because, as the number user increases the orthogonality in walsh codes cannot be maintained.

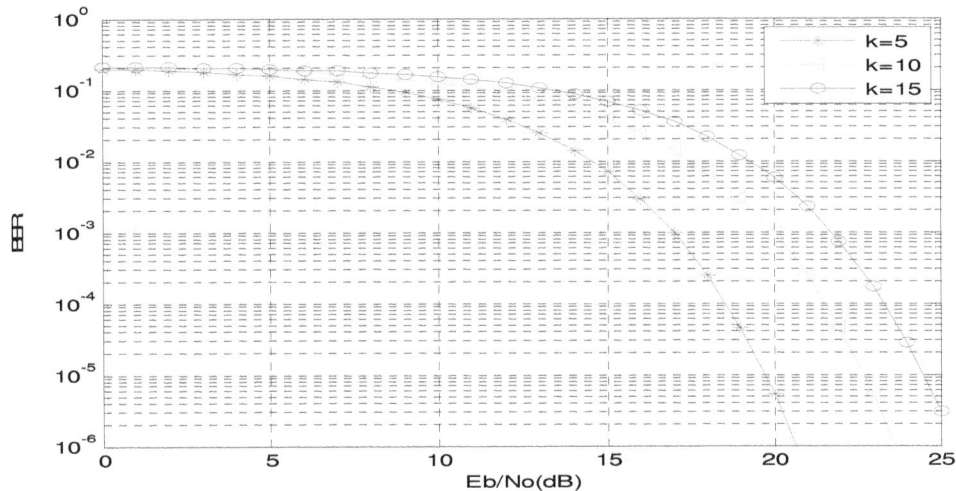

Figure 7. BER analysis under varying Spreading factors

BER performance of CS-QCSK modulation scheme for various spreading factors (K=5, K=10, K=15) are plotted. Figure 7. shows that, as the K value increases the noise performance become worse. In general, smaller Spreading factor corresponds to minimum noise level.

5. CONCLUSION

In the field of chaos-based communication, M-ary DCSK scheme seems to be the best known modulation scheme. But, the drawback is, it requires (N-1) number of delay lines both in the transmitter and the receiver units. In this paper, an improved chaos-based communication method using CS-QCSK technique is recommended for a multiuser scenario. The proposed MU- CS-QCSK scheme, avoids the use of delay lines in the receiver units. It transmits the reference and information signals in the same time slot and offers an increase in data rate, better BER performance and bandwidth efficiency. However, the proposed logic suffers from a marginal increased system complexity. The performance of the proposed modulation scheme is analyzed in a multipath fading channel. Analytical expressions for the BER of MU- CS-QCSK is derived and verified with matlab simulation. Simulation results, shows that the MU-CS-QCSK has better BER performance than the other chaos based multiuser modulation schemes. This proves the suitability of the proposed modulation technique for a mobile radio scenario, which is the need of the hour.

6. REFERENCES

[1]   Abel and W. Schwarz, "Chaos communications—Principles, schemes, and system analysis," *Proc. IEEE*,vol. 90,pp. 691–710, May 2002

[2]   G. Kolumbán, M. P. Kennedy, Z. Jako, and G. Kis, "Chaotic communications with correlator receivers: Theory andperformance limits," *Proc.IEEE*, vol. 90, pp. 711–732, May 2002.

[3]   H. Dedieu, M. P. Kennedy, and M. Hasler,"Chaos shift keying: Modulation and demodulation of a chaotic carrier using self-synchronizing Chua's circuits," *IEEE Trans. Circuits Syst. II*, vol. 40, pp. 634–642, Oct.1993.

[4]   G. Kolumbán, B. Vizvári,W. Schwarz, and A. Abel, "Differential chaos shift keying: A robust coding for chaos communication," Proc. *Int.Workshop Nonlinear Dynamics of Electronic Systems*, pp. 87–92, June 1996.

[5]   G. Kolumbán, G. Kis, M. P. Kennedy, and Z. Jáko, "FM-DCSK: A new and robust solution to chaos communications," in *Proc., Int. Symp. Nonlinear Theory Appl.*, HI, 1997, pp. 117– 120.

[6]   M. P. Kennedy and G. Kolumbán, "Digital communication using chaos,"in Controlling Chaos and Bifurcation in Engineering Systems, G. Chen, Ed. Boca Raton, FL: CRC, 2000, pp. 477–500.

[7]   G. Kolumbán and M. P. Kennedy, "The role of synchronization in digital communications using chaos—Part III:Performance bounds for correlation receivers," *IEEE Trans. CircuitsSyst. I, Fundamental. Theory Appl.,*vol. 47, no. 12, pp. 1673–1683, Dec. 2000.

[8]   G.Kolumban, Z. Jákó, and M. P.Kennedy, "Enhanced versions of DCSK and FM-DCSK data transmissions systems," in *Proc. IEEE ISCAS'*99,vol. IV, Orlando, FL, May/Jun. 1999, pp. 475–478.

[9]   W. M. Tam, Francis C. M. Lau, and Chi K. Tse, "Analysis of Bit Error Rates for Multiple Access CSK and DCSK Communication Systems*" IEEE Transactions on Circuits and Systems—I:Fundamental Theory and Applications,* Vol. 50, No. 5, May 2003

[10]   M. P. Kennedy, G. Kolumbán, G. Kis, and Z. Jákó, "Performance evaluation of FM-DCSK modulation in multipath environments," *IEEE Trans.Circuits Syst. I, Fundam. Theory Appl.,* Vol. 47, No. 12, pp. 1702–1711, Dec.2000.

[11]   G. Kolumbán, "Theoretical noise performance of correlator-based chaotic communications schemes," *IEEE Trans. Circuits Syst. I, Fundam. Theory Appl.,* vol. 47, no. 12, pp. 1692– 1701, Dec. 2000.

[12]   Z. Galias and G. M. Maggio, "Quadrature chaos-shift keying: Theory and performance analysis," *IEEE Trans.Circuits Syst. I, Fundam.Theory Appl.,* vol. 48, no. 12, pp. 1510–1519, Dec. 2001.

[13]   G. Kolumbán and G. Kis, "Multipath performance of FM-DCSK chaotic communications system," in *Proc. IEEE Int. Symp. Circuits and Systems*, Geneva, Switzerland, May 2000, pp. 433–436.

[14]   S. Mandal and S. Banerjee, "Performance of differential chaos shift keying over multipath fading channels," in *Proc. Indian Nat. Conf. Nonlinear Systems and Dynamics*, Kharagpur, India, Dec. 2003.

[15]   Xia, Y.,Tse,C.K.& Lau, F.C.M, "Performance of differential chaos shift keying digital communication systems over a multipath fading channel with delay spread,"*IEEE Trans. Circuits Syst.-II* 51,680-684 ,2004.

[16]   G. Kolumbán, P. Kennedy, and G. Kis, "Multilevel differential chaos shift keying," *in Proc. Int. Workshop, Nonlinear Dynamics of Electronics Systems*, NDES'97, Moscow, Russia, 1997, pp. 191–196.

[17]   F. C. M. Lau, M. M. Yip, C. K. Tse, and S. F. Hau, "A Multiple-Access Technique for Differential Chaos-Shift Keying" *IEEE Trans.Circuits Syst. I, Fundam.Theory Appl.,* vol. 49, no. 1, January 2002.

[18]   YiWei Zhang, Xubang Shen and Yi Ding, "Design and performance analysis of an FM- QCSK chaotic communication system," *International Conference* on *Wireless Communications, Networking and Mobile Computing*, China, pp: 1-4, Sept. 2006.

[19]   Jiamin Pan, He Zhang, "Design of FM-QACSK Chaotic Communication System" *IEEE trans. Wireless Communications & Signal Processing, 2009.*

[20]   Yao,J.& Lawrance, A.J, " Performance analysis and optimization of multi-user differential chaos-shift keying communication systems," *IEEE Trans.circ. Syst.-I* 53, pp.2075-2091, 2006..

[21]   Weikai Xu, Lin Wang and Guanrong Chen, "Performance of DCSK Cooperative Communication Systems Over Multipath Fading Channels", *IEEE Transactions on Circuits and Systems—I: Regular Papers*, Vol. 58, No. 1, January 2011

[22]   Lin Wang, Xin Min, and Guanrong Chen, "Performance of SIMO FM-DCSK UWB System Based on Chaotic Pulse Cluster Signals", *IEEE Transactions on Circuits and Systems,* 2011

[23]   Hua Yang and Guo-Ping Jiang, "High-Efficiency Differential-Chaos-Shift-Keying Scheme for Chaos-Based Noncoherent Communication",   *IEEE Transactions on Circuits and Systems-II: Express Briefs*, Vol. 59, No. 5, May 2012

[24]   G.Koluban,W.K.Xu and L.Wang, "A Novel differential chaos shift keying modulation scheme," *International journal of Bifurcation and chaos*,2011,Vol.21,No.3,pp799-814

[25]   W. K. Xu and L. Wang, G. Kolumb´an, "A New Data Rate Adaption Communications Scheme for Code-Shifted Differential Chaos Shift Keying Modulation", *International Journal of Bifurcation and   Chaos, 2012*

[26]   K.Thilagam and K.Jayanthi, " A Novel chaos based modulation scheme (CS-QCSK) with improved BER Performance", *CS&IT-Computer Science Conference Proceedings*, CoNeCo 2012, pp-45-59,October 2012.

5

MAC LAYER PERFORMANCE: STUDY AND ANALYSIS FOR DIFFERENT MOBILITY CONDITIONS IN WSN

Manjusha Pandey [1], Shekhar Verma [2]

Indian Institute of Information Technology Allahabad, India
{rs58[1], sverma[2]}@iiita.ac.in

ABSTRACT

Wireless sensor networks are the most challenging networks for communication because of its resource constrained nature and the dynamical nature of network topographic anatomy. A lot of research is being going on in the diverse parts of the world for optimum utilization of communication resources in these special types of ad hoc networks. The utility and application domain of sensor networks ranges from commercial, public safety applications and military sector to be the most important ones. The magnificent challenges to the routing algorithms employed in such type of networks are due to the mercurial size of the network and its expandable topology that is quite dynamic in nature. The present paper offers a comparison and analysis of the packet drop at the MAC layer for different routing protocols under an experimental setup having different mobility condition based scenarios of the wireless sensor network application. The comparative study may have also an impact on the improvement of MAC layer performance for different simulation times of the experimental setup considering two of reactive as well as proactive protocols that are most widely used routing protocol in wireless sensor networks. Wireless sensor network application under consideration for the experimental is the battle field monitoring wireless sensor network and the comparative study has been performed for four different mobility patterns described as four different scenarios in the considered experimental application of wireless sensor networks.

The sensor network simulative electronic deception architecture used is for the battle field monitoring application of wireless sensor networks. The application provides support for sensing capabilities within the network nodes called as UGS (Unattended Ground Sensors).Mobile nodes gather data from battle field and direct it to the base station via mobile UGV (Unmanned Ground Vehicles).The performance of the MAC Layer varies with the different average jitter values for different simulation times in the network. Power usage model has been used to reliably represent an actual sensor hardware and sensor network oriented traffic pattern.

KEYWORDS

Medium Access Control Protocol, Wireless Sensor Networks, Packet Drop .AODV, OLSR, DYMO, LANMAR

1. INTRODUCTION

Wireless sensor networking is an emerging technology that has potential usage in environment monitoring, defense, smart spaces, scientific application, medical systems and robotic exploration, target tracking, intrusion detection, wildlife habitat monitoring, climate control and disaster management etc. Wireless sensor network (WSN) consist of one or more battery-operated sensor devices with embedded processor, small memory and low power radio. Low power capacities of sensor node results in limited coverage and communication range for sensor nodes compared to other mobile devices. Hence, to successfully cover the target area, sensor networks are composed of large number of nodes. Nodes in wireless sensor network coordinate to perform a common task.

 Medium access control (MAC) protocol in wireless networks has an important role to enable the successful operation of the network. One fundamental task of the MAC protocol is to avoid collisions so that two interfering nodes do not transmit at the same time. There are many MAC protocols that have been developed for wireless networks. Typical examples include the time division multiple access (TDMA), code division multiple access (CDMA), and contention-based protocols like IEEE 802.11.In a wireless network, controlling when to send a packet and when to listen for a packet are two most important operations to be performed by the medium access layer. In general, idly waiting wastes huge amounts of energy during communication. Medium access control protocol deals with when and how to access medium by a node, and how to transfer the data safely when there is more than one node accessing a single wireless channel simultaneously. MAC layer is a part of DLC (data link layer) which is divided into MAC and LLC (Logical link layer) sub layers. The main task of LLC is Error and Flow Control. MAC layer resolves contentions in a multi-access wireless environment. Problems in medium access are influenced by a number of attributes and trade off's like - Collision Avoidance, Energy Efficiency, Scalability and adaptivity, Channel Utilization, Latency, Throughput and Fairness.

Because of the mobility of nodes, there are certain characteristics that are only applicable to adhoc networks. Some of these key characteristics are bandwidth constrained links, dynamic network topologies, and energy constrained operations. In the real-time applications, and real-time data, the ad hoc networks allow for Quality of Service (QoS) in terms of delay, bandwidth, as well as packet loss. This network does not have defined routers and routes. All nodes have capability of moving, may work as routers, and can be connected in an arbitrary manner. Functioning as routers, these nodes discover as well as maintain routes to other node within the network. The nodes move around randomly, thus making the network topology dynamic in nature. So it is important for the routing protocols to be adaptive and have the ability to maintain routes in spite of dynamic network topology. These networks have drew in a lot of attention throughout the past several years because of increased demand for ubiquitous connectivity and emergence of new communication scenarios such as sensor networks Some critical areas of applications of these networks are in the fields of military and civilian application such as communication in the battle field, disaster management, vehicular movement or communication in traffic management and scientific exploration etc. In all these applications, group communication is more important.

The present research effort concentrates on the sensor network simulator architecture that furnishes support for sensing potentialities in network nodes, existent sensor hardware and sensor network orienting traffic model. We have contemplated sensor network models in the various circumstance of network simulation and this is the exclusive work to our cognition that compares the routing mechanisms for detailed models on the operation of sensor networks [2]. Four different mobility condition based scenarios illustrate the comparative analysis of using accurate and representative wireless sensor network models.

2. STATE OF ART

In wireless ad hoc networks [9] out of numerous views to be taken into thoughtfulness one of the most significant is that of the effective energy management with the additional goal of prolonged connectivity of the network and increased lifetime of the network. These constraints are particularly true of sensor networks. In these networks the nodes are usually battery powered and left unattended after deployment. The routing algorithms designed for these networks need to monitor the energy of nodes and route packets accordingly. Considering the work done in the field of comparison and analysis, the analysis has been done between the routing protocols evaluated based on quantitative and qualitative metrics [14]. But the analysis of a protocols performance for exhaustive variations in simulation time of the same network and for different application scenarios has not been proposed and performed yet. A great deal of research in the domain of routing protocols in ad hoc networks has been done; AODV, DYMO, OLSR, LANMAR to mentioned a few.

OLSR [3] is a variation of traditional link state routing, modified for improved operation in ad hoc networks. The key feature of OLSR protocol is that it uses multipoint relays (MPRs) to reduce the overhead of network floods and size of link state updates. The OLSR protocol executes the hop by hop routing i.e. each node uses its most recent information to route a packet. States involved in the same are as neighbor sensing, multipoint relay station, MPR information declaration, routing table calculation.

LANMAR [12] aggregates the characteristics of Fisheye State Routing (FSR) [13] and Landmark routing. The fundamental novel characteristic is the role of landmarks for each set of nodes that move like a group (e.g., a team of co-workers at a convening or a tank battalion in the battleground) in order to subdue routing update operating expense. On the other hand, On-demand routing protocols like AODV [5], DYMO [8] etc. are more dynamic. Instead of periodically updating the routing information, these protocols update routing information whenever a routing is required. This type of routing produces routes only when in demand by the source node and therefore, in general, the signaling overhead is reduced compared to proactive approaches of routing.

DYMO is meant for use by moving nodes in wireless, multi-hop networks. DYMO determines unicast amongst DYMO routers in the network in an on-demand manner, offering bettered convergence in dynamic topologies. The introductory procedures of the DYMO protocol are route finding (by route request and route reply) and route maintenance. It is an improvement to AODV and more comfortable to implement. In networks with a prominent number of routers, it is best suited for sparse traffic scenarios. In each DYMO router, minimal state routing is preserved and therefore it is relevant to memory constrained devices. The protocol is suitable for scalability. However, it is yet to be explored for its functionality.

3. MOBILITY CONDITIONS AND IMPLEMENTATIONS

This scenario under consideration for the present research effort demonstrates data collection from ground sensors using mobile vehicles for a battle field monitoring system. Sensors are randomly deployed in an observation region. The sensors constantly monitor any phenomena of interest in the area. The sensory information observed by each sensor is stored locally at the sensor. The mobile vehicles have short range communication to sensors and long distance communication to a remote site which is called fusion centre in this scenario. The sensors send their locally stored data packets to the vehicles which at any time are within their radio range. The vehicles then relay sensory data packets to fusion centre using long distance communication to that centre. The sensors which have CBR flows to fusion centre then are able to send their sensory data to the centre. The four different conditions for the scenario taken as mentioned above may be described as follows:

a) All nodes of the network are static. The UGS (unattended ground sensors) as well as the UGV (Unmanned ground station) are static while the fusion centre remains static in each and every condition in which the scenario has been implemented.

b) In the second condition of scenario implementation the UGS (unattended ground stations) are static while the UGV (unmanned ground station) are mobile.

c) The third condition of scenario says the implementation of the UGS (unattended ground stations) is mobile while the UGV (unmanned ground station) are static.

d) The last condition of scenario implementation refers to situation when the UGS (unattended ground stations) are mobile as well as the UGV (unmanned ground station) are also mobile.

4. METRIC AND METHODOLOGY OVERVIEW

Advantages of MAC protocols for sensor networks include that the energy waste caused by idle listening is reduced by sleep schedules. Beside implementation simplicity, global time synchronization overhead may be prevented with sleep schedule announcements. There are also some disadvantages of MAC protocols for sensor network are, having a fixed duty cycle i.e. Active time is fixed. It is not optimal. If message rate is less energy is still wasted in idle-listening. Sleep and listen periods are predefined and constant which decreases the efficiency of the algorithm under variable traffic load. Long listening interval is expensive - Everyone stays awake unless somebody transmits .Time synchronization overhead even when network is idle.RTS/CTS and ACK overhead when sending data.

While traditional MAC protocols are designed to maximize packet throughput, minimize latency and provide fairness, protocol design for wireless sensor networks focuses on minimizing energy consumption. The application determines the requirements for the minimum through-put and maximum latency. Fairness is usually not an issue, since the nodes in a wireless sensor network are typically part of a single application and work together for a common purpose. The major sources of energy waste in a MAC protocol for wireless sensor networks are the following:

a) Collision: When a transmitted packet is corrupted it has to be discarded, and the follow-on retransmissions increase energy consumption.

b) Control Packet Overhead: Sending and receiving control packets consumes energy too, and less useful data packets can be transmitted.

c) Idle Listening: Listening to receive possible traffic that is not sent can consume extra energy.

d) Overhearing: Meaning that a node picks up packets that are destined to other nodes can unnecessarily consume energy.

The medium-access layer has two functions that impact packet delivery performance: arbitrating access to the channel, and (optionally) some simple form of error detection. In addition to factors that impact the physical layer, and hence the performance of medium-access, two factors affect the medium-access layer. First, the application workload (and, in the case of sensor networks, the sensed environment) determines the traffic generated by nodes and hence the efficacy of channel access. Second, the topology (or, equivalently, the spatial relationship between nodes) affects how many nodes might potentially contend for the channel at a given point in time. To understand packet delivery performance as observed at the MAC layer, we use the following general setup. We place sixty nodes in a somewhat ad-hoc fashion, but at densities that we expect of sensor network deployments. Each node periodically generates a message destined to one of its neighbors; the periodicity of this message generation defines an artificial workload. We then place this setup in three environments as before, and measure several aspects of packet delivery performance.

The overall goal of our experiments was to compare and analyze the packets dropped ratio of the two considered reactive and proactive protocols for various application scenarios considered for the experimental simulations. Also the analysis has been done for routing protocol during the variations in the simulation time for the experimental setup. The protocols were carefully implemented according to its specifications. During the process of implementation of the AODV routing protocol and analyzing the results for each simulation runs, we discovered some modifications in the average jitter of the network for each simulation interval the network varied its performance, while carrying on to succeed to deliver data packets to their destinations. To measure these variations, our basic methodology was to apply to a simulated network a variety of, simulation intervals and different application scenarios implementing various mobility conditions that affect the routing protocols performance, and it's testing with each data packet originated by some sender whether the routing protocol can at that time route to the destination of that packet. We were not attempting to measure the protocols' performance on a particular workload taken from real life, but rather to measure the protocols' performance under a range of conditions.

5. APPLICATION AND ANALYSIS

The average delivery ratio decrements as channel error rate gains due to the increased packet loss error rate and this begins, reflecting the packet loss obtained both by increased congestion and due to packet loss at the MAC Layer. The packet loss rate at the MAC layer (between two routers, or between a router and a host) must be made very small in order to achieve better network routing protocol performance .It is the job of the MAC layer to achieve this condition of optimized data packet control.

The primary aspect of wireless communication performance of interest to us is packet delivery performance. More precisely, our primary measure of performance is packet loss rate (the fraction of packets that were transmitted within a time window, but not received) or its complement, the reception rate. There are many, many factors that govern the packet delivery performance in a wireless communication system: the environment, the network topology, the traffic patterns and, by extension, the actual physical phenomena that trigger node communication activity. It is difficult to isolate these phenomena in order to study the impact of different factors on packet delivery performance. Rather, we take a somewhat mechanistic view in this paper, and look at the packet delivery performance at two different layers in the networking stack: the physical layer and the medium-access layer

AODV routing protocol with the increase in the simulation time represents a steep variation from a high rate of packet loss for the highest simulation time having the highest average of 2700 packets lost to a low of 100 packets lost during the lowest simulation interval for which the application was implemented Though for the lesser variation in the simulation time the net packet loss at the MAC layer does not varies much.

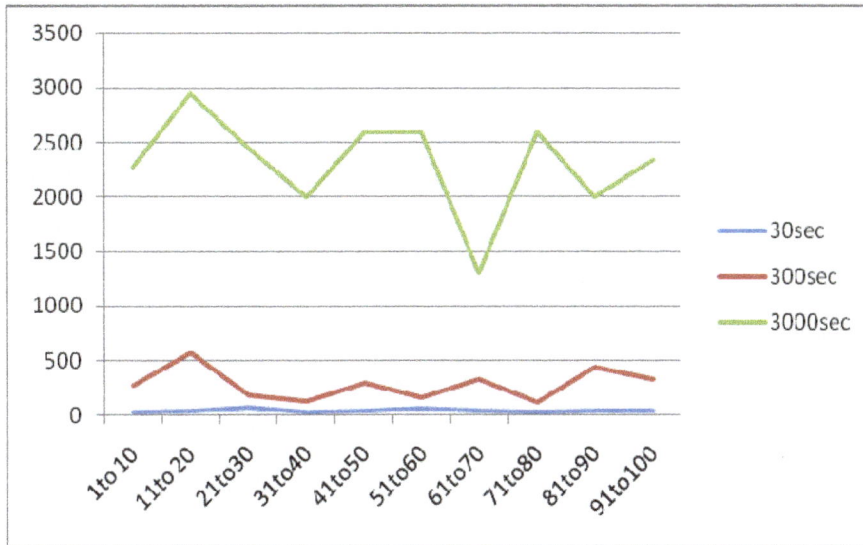

Figure 1. Packet Drop at MAC layer for AODV.

DYMO being an on demand reactive routing protocol like AODV still performs best for higher simulation time while the net packet loss presenting a minimum value of less than 100 packets loosed for the highest simulation time though with decrease in the simulation time the packet drop rate has increased thus making DYMO a suitable choice for longer period of simulation or network utilization in case of real life applications.

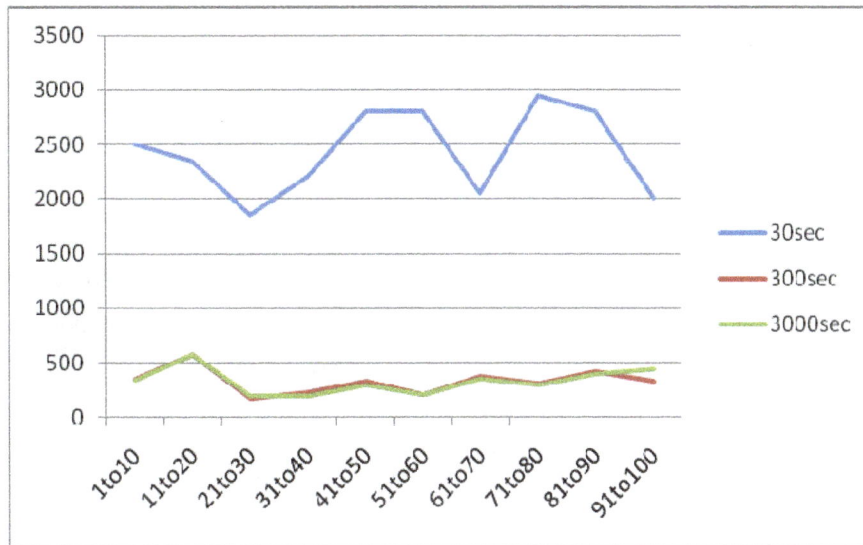

Figure 2. Packet Drop at MAC layer for DYMO.

OLSR uses the proactive methodology of routing techniques depending on the link states of the network. The results being analyzed show that this routing protocol produces best MAC layer performance in the case of smallest simulation time interval with the lowest number of packets dropped in that case being lesser than 50 packets dropped and the maximum packet drop was visible for the largest simulation time the highest value of packet dropped being 4500.

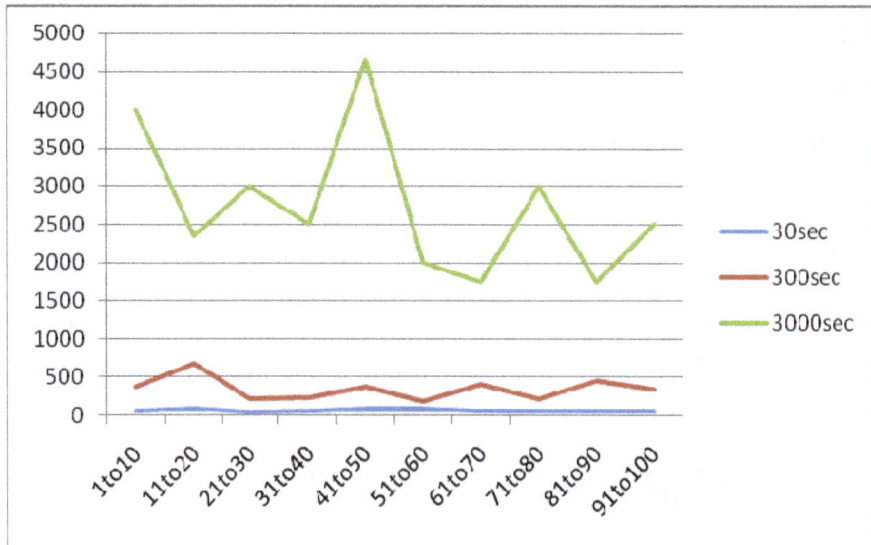

Figure 3. Packet Drop at MAC layer for OLSR.

LANMAR that is also a proactive routing protocol depicts the similar results as the earlier proactive routing protocol and of performs best in the case of smallest simulation time that is taken to be 3000 seconds. The highest packet drop witnessed at the MAC layer in this case of routing is about 3500 packets and the lowest being above 50 packets.

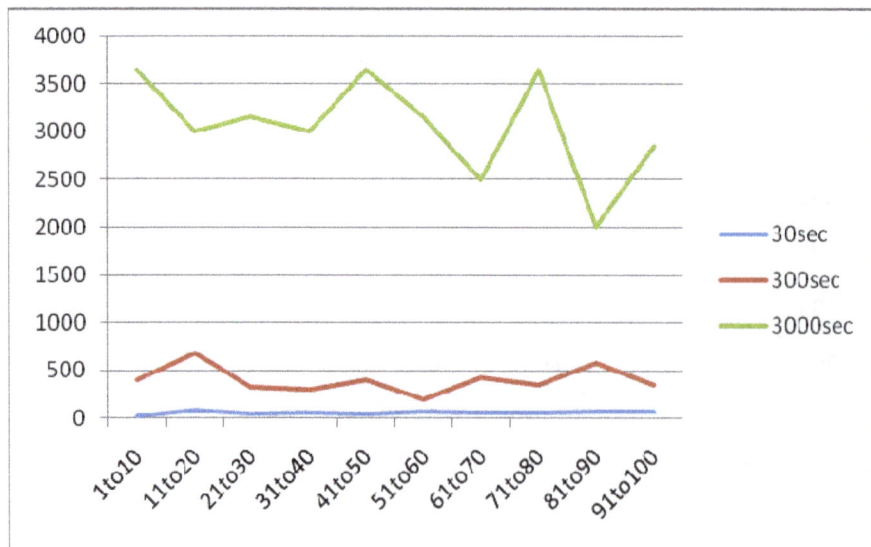

Figure 4. Packet Drop at MAC layer for LANMAR

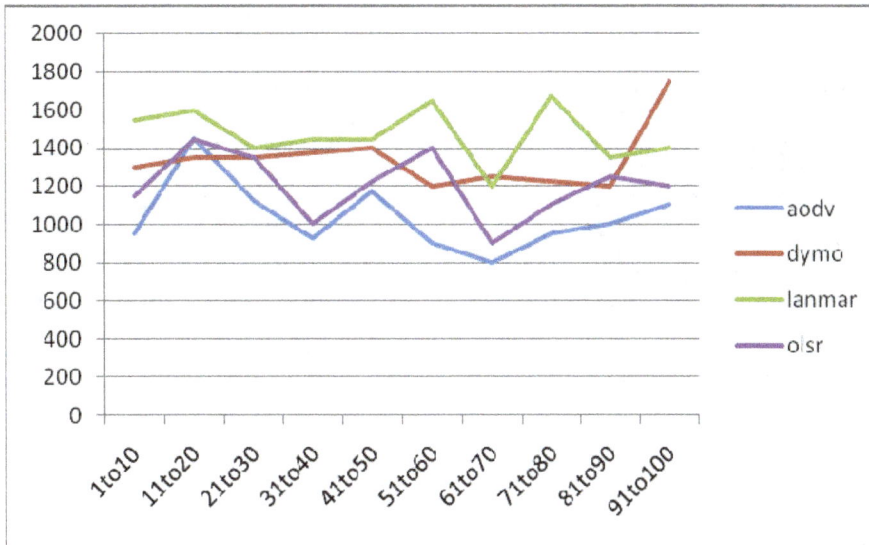

Figure 5. Packet Drop at MAC layer for First Scenario.

For the first scenario having the UGSs as well as the UGVs static the OLSR performs the best with the lowest number of packets dropped at the MAC layer and having the least high packet dropped value of 2900, and LANMAR being the least performing routing protocol with the highest packet drop rate of more than 5000.

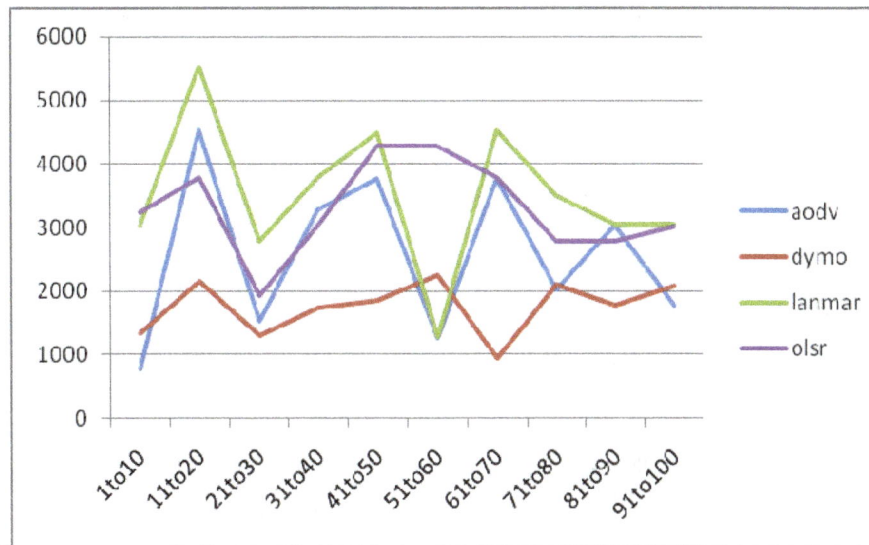

Figure 6. Packet Drop at MAC layer for Second Scenario.

In the case of second scenario having the UGSs as Static and UGVs mobile the total number of packets dropped is highest for the OLSR routing protocol having the packet drop rate more than 2900 and the best MAC Layer performance is depicted by the AODV having the lowest packet drop rate as below as 500 packets.

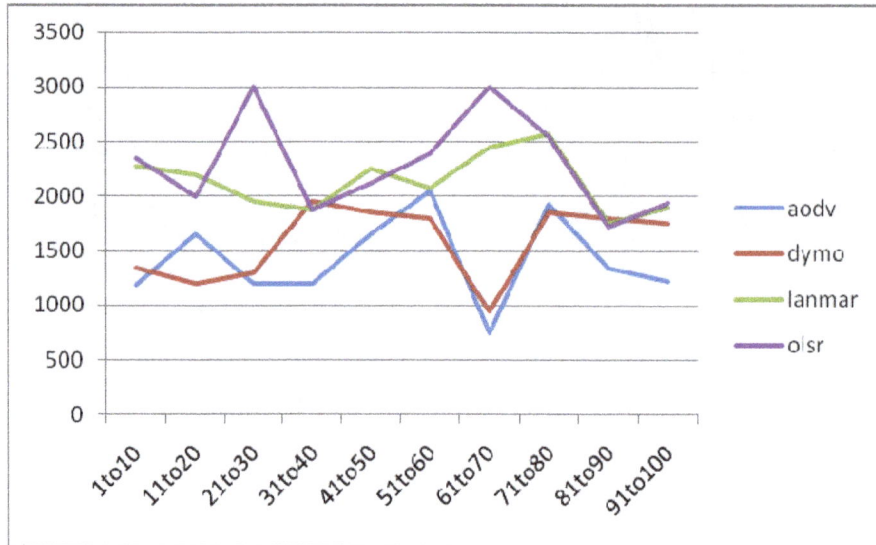

Figure 7. Packet Drop at MAC layer for Third Scenario.

For the third scenario having the UGSs as mobile and UGVs static the lowest packet drop rate has been depicted by the AODV protocol having average packet as low as 800 packets .While the highest packet drop rate has been witnessed in the case of OLSR having the highest packet drop rate of more than 2900 packets being dropped.

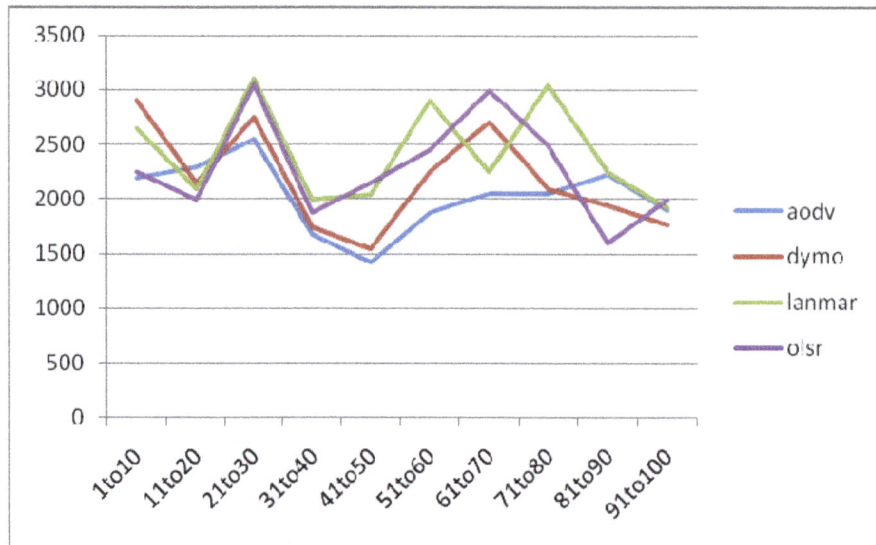

Figure 8. Packet Drop at MAC layer for Fourth Scenario

In the case of fourth scenario having both the UGSs as well as UGVs are mobile the AODV depicts the best performance with the lowest number of packets being dropped having the lowest packet drop value of 700 packets .While the worst performance is being depicted by the DYMO routing protocol in this scenario having the highest packet dropped value of 1700.

Recent surveys in sensor network simulation can be categorized between less flexible but precise simulation based advance and more generic but less elaborated network simulator models. Simulator which furnishes a rich suite of following models: sensing stack to model wave and diffusion anchored sensor channels, a precise battery model, processor power use model, energy usage model and sensor network based traffic pattern. We also introduce our study on the effects of elaborated modeling on the functioning of higher layer protocols. We describe the affect of using precise models for battery, processor power usage and tracking models on the network layer stats as network lifetime and accessibility, throughput and routing operating cost. Our results show that comparative MAC layer packet drops at various time durations of simulation. Next section discusses the results for packets dropped at the MAC Layer when the two reactive and proactive protocols were implemented for various mobility condition based scenarios.

6. CONCLUSION

Designing a MAC protocol which can improve energy-efficiency to extend network lifetime in wireless sensor networks is a challenging problem. It is mainly due to stringent resource constraint both in sensor nodes and in wireless media. Several energy-efficient medium access control protocols both contention-based and reservation-based for the wireless sensor network that have been proposed by the researchers are presented in this paper. The design of an optimized MAC protocol for energy efficiency also depended on the actual application. However, no specific MAC protocol has been accepted as a standard. Another reason is the lack of standardization at lower layers (physical layer) and the sensor hardware. Therefore, it will be difficult to have one standard MAC protocol which will work for all possible WSN applications. Therefore, still a lot of work has to done in working out a MAC protocol which will adapt its behavior based on the applications. The research area of ad hoc and sensor network has very much attracted the academic domain as well as the industry both due to its wide-ranging possible application for anytime, anywhere, and any how communication scenarios.

This wide spectrum of applications possible for sensor networks has made the network vividly applicable and acceptable. The routing protocol for sensor networks has been a dynamic research area altogether through the present decade. Although wide efforts have been exercised so far on the routing problem in wireless communications, there are still some challenges via multicasting that still confront effective solutions to the routing problem. A number of such protocols have been purposed developed and implemented also. But no protocol has been found to be best for the wide domain of sensor network applications. Each protocol possesses its advantages and disadvantages. Counting the constraints followed by the networks the routing algorithms have been updated and modified time to time to make the routing more and more efficient and accurate. The present work proposes to find out the effect of different patterns of node mobility within the network. The results though don't present a steep comparative orientation of the results towards a specific routing protocol but the comparative study leads towards some interesting results.

Further research is needed to find most suitable protocol for each and every scenario condition so that an optimized routing protocol could be suggested for various real life applications have concurrency to the mentioned scenarios of the simulated wireless network environment.

REFERENCES

[1]   C.K Toh, & E.M.Royer, (Apr, 1999) "A review of current routing   protocols for ad hoc mobile wireless network", *IEEE Personal Communications* 15, pp 46-55.

[2]   I.F Akyildiz, Su. Weilian, Y.Sankarasubramaniam, & E Cayirci, (2002) "A survey on sensor networks" *Communications Magazine, IEEE Volume: 40 Issue:8* , pp 102-1148.

[3]   C.Santivanez, B.,McDonald. I.,Stavrakakis,& R.Ramanathan "Making link state routing scale for ad hoc network". DOI=10.1145/501417.501420.

[4]   S Basagni, M.Conti, S.Giordano, & I.Stojmenovic (2004) "Mobile Ad Hoc Networking" *IEEE Press and John Wiley and Sons, Inc., New York.* ISBN 0-471-373133.

[5]   C.Perkins, E.B.Royer, & S Das, (2003) "Ad hoc on-demand distance vector (AODV) routing" RFC 3561.

[6]   I.Chakeres, & C.Perkins, (2008) "Dynamic MANET On-demand (DYMO) Routing". IETF MANET, Internet-Draft, 5 December , draft-ietf-manet-dymo-16.

[7]   Qualnet: http://www.scalablenetworks.com.

[8]   W.,B Heinzeelman, A.Chandra Kasan, , & H. Bala Krishnan *(2000)* "Energy-Efficient communication protocol for wireless micro-sensor networks." *Proceedings of the 33rd Hawaii International Conference on System Sciences* , pp. 3005–3014.

[9]   C.Siva Rama, Murthy& , B.S Manoj "Ad hoc Wireless Networks: Architecture and protocols,"*Second edition, Prentice Hall.*

[10]   Sanatam Mohanty, & Sarat Kumar Patra, (2010) "Performance Analysis of Quality of Service Parameters for IEEE 802.15.4 Star Topology using MANET routing" *ACM 978-1-60558-812-4.*

[11]   Maneesh Varshney, & Rajiv Bagrodia (2004) "Detailed models of sensor networks simulations and their impact on network performance" *Proceeding of MSWiM '04 Proceedings of the 7th ACM international symposium on Modeling, analysis and simulation of wireless and mobile systems.*

[12]   P. F. Tsuchiya, (1988) "The Landmark Hierarchy: a new hierarchy for routing in very large networks", *In   Proceeding   of   SIGCOMM   '88   Symposium   proceedings   on Communications architectures and protocols Issue 4 vol.18*, pp 35-42.

[13]   H. Ren, M. Q H.,-Meng, & X Chen *(2006).* "Investigating network optimization approaches in wireless sensor networks". *In proceedings of International conference on Intelligent Robots and Systems, IEEE/RSJ*, pp2015—2021.

[14]   S.R. Das, R. Castaneda, J.Yan, & R. Sengupta, (1998) "Comparative performance evaluation of routing protocols for mobile ad hoc networks" *in Proceedings of 7th international Conference on Computer Communications and Networks,* pp153–161.

6

Effective Capacity of a Rayleigh Fading Channel in the Presence of Interference

Mehdi Vasef

University of Duisburg-Essen, Germany

`mehdi.vasef@stud.uni-due.de`

Abstract

In recent years the concept of the effective capacity that relates the physical layer characteristics of a wireless channel to the data link layer has gained a lot of attraction in wireless networking research community. The effective capacity is based on Gärtner-Ellis' large deviation theorem and it is used to provide the statistical QoS provisioning in the wireless networks. Effective capacity also helps in the analysis of the resource allocation or scheduling policies in various wireless systems such as Relay networks, multi-user systems and multi-carrier systems subject to statistical QoS requirements. The effective capacity in noise limited wireless network has already been investigated in the recent works. Considering the interference limited wireless channels, in this paper we propose an analytical approach based on Laplace's method for the effective capacity of uncorrelated Rayleigh fading channel in the presence of uncorrelated Rayleigh fading interference. The accuracy of the analytical model for the effective capacity is validated by numerical simulations. We also provide the evaluation of tail probability of the delay and maximum sustainable rate. The validation results reveal that the proposed mathematical approach to the effective capacity can open the path for further researches in statistical QoS provisioning in interference limited wireless networks.

Index Terms : Cross Layer Optimization, Statistical Quality of Service Provisioning, Large Deviation Principle, Gärtner-Ellis Theorem, Delay Probability, Effective Capacity

I. INTRODUCTION

The theory of statistical QoS (Quality of Service) provisioning has been developed in the early 90' to study wired asynchronous transfer mode (ATM) networks. This theory analyses network statistics such as buffer overflow probabilities, queue distributions and delay-bound violation probability. As a part of the statistical QoS theory, effective capacity (EC) is a promising technique for analyzing the statistical QoS performance of wireless networks where the service process is regarded as a time-varying wireless channel. Effective capacity was initiated by Wu and Negi in their paper [1]. After that EC has been applied in various fields of wireless networks and systems for instance in cognitive radio, downlink scheduling in cellular networks, relay networks, multiple antennas systems, QoS aware routing, multimedia transmission, multi-user scheduling , multi casting and finally performance evaluation in multi hop networks. EC based methods help us generally in the efficient design of upper layer protocols such as channel dependent scheduling algorithms, admission control, channel coding algorithms, adaptive modulation and finding optimal power allocation strategies.

Wu and Negi [5] obtain the EC of for correlated Rayleigh fading channels. They used spectral-estimation-based algorithm to estimate the EC function. In [18], the authors propose a EC model to characterize the Nakagami-m fading channel for QoS provisioning over wireless networks. In [4], the EC technique was used to derive QoS measures for general scenarios, i.e.

multiple wireless links, variable bit rate arrivals instead of constant bit rate, packetized traffic and wireless channels with some propagation delay. In [19], the authors extend the EC model for a single-hop wireless link, to analyse the delay performance of multi-hop wireless link in wireless mesh networks. A fluid traffic model with cross traffic and a Rayleigh fading channel are considered in their study to calculate the end to end delay bound.

[20], [13] discuss about the application of EC in cognitive radios. [20] derives the optimal rate and power adaptation strategies that maximize the EC in cognitive radio networks with uncorrelated Nakagami-m fading. The authors consider maximizing the throughput of spectrum-sharing systems while satisfying the delay QoS constraint.

In [13] the EC is used to quantify the relationship between channel sensing duration, detection threshold, detection and false alarm probabilities, QoS metrics and fixed transmission rates.

[14], [21], [25] explore the EC usage in MIMO (Multiple Input Multiple Output) systems. The article [14] explores the EC of a class of MIMO systems with Rayleigh flat fading. The performance of the EC for MIMO systems assuming the large number of transmitting and receiving antennas. A lower bound on the EC is obtained for the general multi antenna systems in the large antenna array regime. Jorswieck et al [21] obtain the optimal transmit strategy that maximize the EC of the MIMO link with covariance feedback. They also characterize the optimal power allocation and beam-forming optimality range as function of QoS exponent and SNR. Furthermore, they show that increasing the QoS exponent decreases the beam-forming optimality range. Gursoy [25] studies the performance of MIMO systems working under statistical QoS constraints. He uses the EC as performance metric that provides a throughput under such constraints. They investigate the EC in low power, wide band an high SNR regimes.

[24], [26] investigate the EC in multi-user wireless systems. [6], [27] study the EC application in relay networks. Tang and Zhang[6] focus on two relay protocols namely, amplify-and-forward (AF) and decode-and forward (DF), and develop the associated dynamic resource allocation algorithms, where the resource allocation policies are functions of both the network channel state information (CSI) and the QoS exponent. The resulting resource allocation policy provides a guideline on how to design the relay protocol that can efficiently support stringent QoS constraints. Shaolei and Lataief [27] an algorithm for time-slot allocation which increases the EC for wireless cooperative relay networks in either uncorrelated or correlated fading channels.

[16], [15] apply the EC in downlink wireless networks. [15] obtains the EC of a downlink scheduling system under statistical QoS constraints. The authors derive a generalized class of the optimal scheduling schemes for achieving the boundary points of the capacity regions for two users.

Soret et al [2] study the EC of the noise limited channel for uncorrelated Rayleigh fading and correlated Rayleigh fading. They assume that instantaneous service rates are uncorrelated and according to the central limit theorem, the cumulative service rate can be approximated with a normal distribution. Based on this assumption, they compute the EC function in both uncorrelated and correlated Rayleigh fading cases. The also discuss the derivation of EC for the discrete and continuous transmission cases. Furthermore they obtain the tail distribution of target delay probability and also maximum sustainable rate that the wireless system can support to meet the delay probability.

None of the aforementioned works address the computation of the EC for interference limited channels not only from the theoretical point of view but also empirically. Among these works, the Soret et al's work [2] is much related to our work in the sense that they acquired the EC for the noise limited uncorrelated and time correlated Rayleigh faded channels.

Our experimental results reveal that we cannot treat the summation of interference and noise in the interference limited channels as a noise with large power and just follow the Soret's [2] results. So it is necessary to have precise formulas for EC to get rid of this engineering pitfall and perform the statistical QoS in wireless networks accurately and efficiently. In this paper we propose an analytical approach to obtain the effective capacity in the interference limited wireless networks. To best of our knowledge we are the first who compute the effective capacity for uncorrelated Rayleigh fading channel in the presence of a single uncorrelated Rayleigh faded interferer.

The reminder of the paper is organized as follows. In section II we explain the system model. Then in section III we present the analytical result for the effective capacity in the interference limited channel. Section IV discusses about the validation and also the evaluation of the delay probability and also maximum sustainable rate. Finally section V concludes the paper.

II. SYSTEM MODEL

In the following we first present the system model. We consider a wireless communication with one transmitter and one receiver in the presence of one interfering transmitter. Without loss of the generality, in this paper we do not focus on special wireless communication standard. This simple wireless communication setting can be used as a basis to build via bootstrapping more complex wireless systems such as cognitive radio networks with one primary users and multiple secondary users, relay networks, MIMO systems, multi-carrier systems and a static wireless multi hop networks with arbitrary topology.

The main purpose of traffic modelling is to build models that characterize the statistical properties of data collected through traffic monitoring tools. Recent studies have shown that today's data traffic in wired and wireless network 3 has some properties such has long range dependence (LRD) and heavy-tailedness that traditional models such such Poisson or constant bit rate (CBR) models cannot explain them. For understanding the definition of LRD and its application in modelling data traffic the interested reader can refer to [9].

Due to mobility of the nodes, unpredictable channel behavior and unreliability issues, modelling traffic in wireless networks is more difficult than wired networks. There are some contributions discussing the end to end delay in wireless networks in the presence of LRD traffic. Pezaros et al's paper [8] is one of the recent papers that investigate the relationship between the Hurst parameter and packet delay in a empirical manner.

The arrival is denoted by $a[i]$ and the cumulative arrival rate is represented by $A[i] = \sum_{j=0}^{i-1} a[j]$. Regarding the latest progress in the traffic models for the wireless network, the traffic model that we consider in this paper is CBR fluid flow model. So the arrival rate in CBR is simply $a[i] = \lambda$. Note that in fluid flow traffic models the sources are infinitesimal in size. But in packetized traffic the length of packets plays a role in queueing analysis. Even we may be have variable length packets if we have variable bit rate traffic for instance. We would like to emphasize that the primary objective of the paper is to acquire the effective capacity. Therefore the CBR assumption is reasonable.

At the receiver side, we are interested to evaluate certain QoS requirements given by Pr $(D > D_t)$ and D_t. Pr $(D > D_t)$ represents delay violation probability or simply delay probability, i.e.

the probability that the stationary end to end delay exceeds some certain target delay. D_t shows the target delay.

The channel model that is used throughout the paper is the slow varying frequency flat Rayleigh fading model. Slow varying means the coherence time is greater than the duration of the block. We assume that we have a block fading. In block fading channels, the fading component do not change during the coherence time. Block fading channels are also called quasi-static channels. Furthermore we assume that transmitter has perfect knowledge of CSI.

Transmission time T is divide into N timeslots T_i of equal length so that $\sum_{i=1}^{N} T_I = T$. The datas are entered into a queue at each time slot. If there are some datas in the queue, they are fetched from the queue and are transmitted over the wireless channel. Only one transmission occurs at each timeslot and there is no bursty transmission. We assume that we have always successful transmission in each time slot, hence we the probability of successful transmission is 1.

The transmitter and receiver and interfering transmitter are distanced from each other which causes a path loss. The path loss is constant over time slot i. The average signal-of-interest power at the and average interference power at the receiver due to path loss are denoted by P_{TX} and P_I respectively. In different time slots i, the fading component varies. We represent the fading component for the signal-of-Interest and interfering signal with $h_{TX}^2(t)$ and $h_I^2(t)$. Since we consider a Rayleigh fading model, they are exponentially distributed with mean 1. The fading component is exponentially distributed if the signal-of-interest and interfering signal are characterized by the uncorrelated Rayleigh fading.

In the interference limited wireless channels, the SINR (Signal to Interference plus Noise Ratio) is a good metric for evaluating the quality of transmitted signal. SINR depends on the fading model of the signal-of-interest, interfering signal and also noise. The instantaneous SINR or γ_{SINR} is defined as [7]:

$$\gamma_{SINR}[i] = \frac{P_{TX} \cdot h_{Tx}^2[i]}{P_I \cdot h_I^2[i] + \eta^2} \tag{1}$$

where P_{TX} and P_I are fixed over all t and $h_{TX}^2(t)$ and $h_I^2(t)$ represent the random fading component of the signal-of-interest and of the interfering signal at time slot i. The instantaneous channel capacity that is known as Shannon's capacity is expressed as:

$$C[i] = B_w \log_2 \left(1 + \Gamma \gamma_{SINR}[i]\right) \tag{2}$$

where B_w is bandwidth and Γ is called SINR gap [10] that shows the reduction in SINR with respect to capacity. It depends only on Bit Error Rate (BER). SINR Gap is a quantitative way to mention that we transmit below the Shannon's capacity practically and the channel is under some erroneous condition. We assume that the $B_w = 1khz$ and BER is zero, so $\Gamma = 1$ and the instantaneous channel capacity is simplified as:

$$C[i] = B_w \log_2 \left(1 + \gamma_{SINR}[i]\right) \tag{3}$$

Shannon's capacity model is in fact the theoretical upper bound for the throughput that a wireless channel can achieve for a given SINR or SNR. We do not deal with special modulation or demodulation scheme such as adaptive modulation.

The service process in wireless channels has a time varying nature and is modelled by a Shannon's capacity i.e. $S[i] = C[i]$. The instantaneous service rate is represented by $s[i]$ and the cumulative service rate is defined as $S[i] = \sum_{j=0}^{i} s[j]$ and also we have $S[1] = 0$. The asymptotic effective capacity for the cumulative service rate is:

$$\alpha_S(u) = \lim_{N \to \infty} \frac{\ln E\left(e^{-(S[N] - S[1])u}\right)}{Nu} = \lim_{N \to \infty} \frac{\ln E\left(e^{-(\sum_{i=1}^{N} s[i])u}\right)}{Nu} \tag{4}$$

$$\alpha_S(u) = \lim_{i \to \infty} \frac{\Lambda_{S[N]}(-u)}{Nu} = \frac{N\Lambda_{s[i]}(-u)}{Nu} \tag{5}$$

$$\alpha_S(u) = \frac{\Lambda_{s[i]}(-u)}{u} \tag{6}$$

The cumulative service rate is regarded as a random variable with some probability distribution function (PDF). It is difficult to acquire the closed form for the PDF of the cumulative service rate. Assuming that instantaneous service rates are ergodic, stationary and uncorrelated. We approximate the distribution of cumulative service rate with a normal distribution. It can be easily shown that the cumulant generating function of the normal distribution is given by [29]:

$$\Lambda_{s[i]}(u) = E[s[i]]u + \text{Var}[s[i]] \frac{u^2}{2} \tag{7}$$

So the effective capacity is given by:

$$\alpha_S(u) = E[s[i]] - \text{Var}[s[i]] \frac{u}{2} \tag{8}$$

To obtain the effective capacity, we need to only calculate the statistical moments of the service rate. In the interference limited channels, the service rate is related to SINR rather than SNR. If the SINR is high, the channel is in a good state and the datas can be quickly served by a queueing system. When the SINR is low, the channel is in a bad state and delay probability will be high.

The single server queue plays a central role in the performance modelling and evaluation of the wired and wireless networks and is the vital element in our system model. We consider a queueing system with a single input and single output with infinite length. We assume that there is no dropping policy. For the sake of simplicity, the scheduling scheme is work conserving first in first out (FIFO). From work conserving we mean that if some datas are in queue, they will be served and the server will never be idle. Certainly there are more sophisticated scheduling policies such as wireless fair queueing [12] and modified largest weighted delay first (M-LWDF) that was proposed in [11]. Wireless fair queueing scheme uses the idea of fair scheduling for wired network in wireless networks. Albeit this scheme provides fairness, but it can not guarantee the stringent QoS requirements. The M-LWDF scheme does not perform the explicit QoS provisioning such as delay probability in terms of the arrival rate.

Even more intricate concepts in queueing theory can influence the statistical QoS provisioning in wireless networks. For example in our queueing system we do not take into account the priority among the traffic sources. Xie and Hengei [17] study the priority queues over Rayleigh fading channels and evaluate the packet loss probability.

We discard the transient queue length and delay in our analytical approach, but also in queueing simulations. Exploring the impact of the two aforementioned scheduling policies on effective capacity and QoS exponent and also finding the optimal time slot length or optimal power allocation scheme are irrelevant to our system model.

The length of queue in terms of the arrival rate λ and the service rate S [i] is:

$$Q[i] = \max\left(0, Q[i-1] + S[i-1] - \lambda T_s\right) \tag{9}$$

Where T_s is the sampling interval. We assume that in each time i the service rate is determined by the Shannon's capacity, i.e. $S[i] = C[i]$ where $C[i]$ is the amount of the Shannon's capacity at time slot i and is given by $C[i] = \sum_{t=t_{b_i}}^{t_{b_i}+T_i} C[t]$. t_{b_i} is the beginning time of time slot i and T_i is the duration of time slot i.

Regarding the CBR model and using the Gärtner-Ellis' large deviation theorem, the tail of queue in steady state is [1], [28]:

$$\Pr(Q > B) \approx \xi e^{-Bu} \tag{10}$$

Where η denotes the probability that the queue is not empty. It is easy to check that $u = \dfrac{2(E[s[i]] - \lambda)}{Var[[s_i]]}$. The tail probability of the delay is given by:

$$\Pr(D > D_t) \approx \xi e^{-D_t \lambda \frac{2(E[s[i]] - \lambda)}{Var[[s_i]]}} \tag{11}$$

The above formula, is a basis for the analysis of the delay probability in both uncorrelated and time correlated service rates. To evaluate the delay probability we need to compute the E[s[i]] and Var [s [i]]. In the next section we will propose analytical approaches to obtain them.

III.   EFFECTIVE CAPACITY OF INTERFERENCE LIMITED WIRELESS NETWORKS

In this section we will obtain the EC in uncorrelated Rayleigh fading case. We first compute the distribution of SINR. Then we continue to obtain the mean of service rate in exact form. We then propose a solutions for the second moment of the service rate based on Laplace's method.

A. Derivation of the SINR in uncorrelated Case

In this section we obtain the distribution of SINR in uncorrelated case using transformation of random variables. First of all, we review a theorem about transformation of the random variables.

Theorem 1: Let $f_{X,Y}(x, y)$ be the value of joint PDF of the continuous random variables X and Y at (x, y). If the functions $u = g_1(x, y)$ and $v = g_2(x, y)$ are partially differentiable with respect to x and y and represent a one-to-one transformation for all values within the range of X and Y such that $f_{X,Y}(x, y) \neq 0$, then for these values the equations can be uniquely solved for x and y to give $x = w_1(u, v)$ and $y = w_2(u, v)$ and for corresponding values of u and v, the joint PDF of $U = g_1(X, Y)$ and $V = g_2(X, Y)$ is given by [29]:

$$f_{U,V}(u, v) = f_{X,Y}(w_1(u, v), w_2(u, v)) \, |J| \tag{12}$$

where $|J|$ is the determinant of Jacobian matrix of the transformation and is defined as:

$$|J| = \begin{vmatrix} \frac{\partial x}{\partial u} & \frac{\partial x}{\partial v} \\ \frac{\partial y}{\partial u} & \frac{\partial y}{\partial v} \end{vmatrix}$$

From section II, we know that SINR is defined as:

$$\gamma_{\text{SINR}}(t) = \frac{X(t)}{Y(t) + \eta^2} \tag{13}$$

The signal-of-interest and interfering signal are exponentially distributed with average powers P_{TX} and P_I respectively. The signal-of-interest and interfering signals can also be expressed in terms of instantaneous power [22]

$$f_X(x) = \frac{1}{P_{TX}} e^{\frac{-x}{P_{TX}}} \tag{14}$$

$$f_Y(y) = \frac{1}{P_I} e^{\frac{-y}{P_I}} \tag{15}$$

where P_{TX} and P_I are average powers of signal-of-interest and interfering signals respectively. We find easily $w_1(u, v) = u(v + \eta^2)$ and $w_2(u, v) = v$. Applying the theorem 1 yields:

$$f_{U,V}(u,v) = f(u(v+\eta^2), v)(v+\eta^2) \tag{16}$$

since the X and Y are independent, so the joint PDF can be decomposed i.e.

$$f_{U,V}(u,v) = f(u(v+\eta^2)), f(v)(v+\eta^2) \tag{17}$$

We are interested to find the distribution of u, so we can integrate the joint PDF of $f_{U,V}(u, v)$ with respect to v.

$$f_U(u) = \int_0^\infty f\left(u\left(v+\eta^2\right)\right) f(v)\left(v+\eta^2\right) dv \tag{18}$$

$$f_U(u) = \frac{1}{P_{TX}\tilde{P}_I} e^{-\frac{u\eta^2}{P_{TX}}} \int_0^\infty e^{\left[-\left(\frac{u}{P_{TX}}+\frac{1}{P_I}\right)v\right]} \left(v+\eta^2\right) dv \tag{19}$$

$$f_U(u) = \frac{1}{P_{TX}\tilde{P}_I} e^{-\frac{u\eta^2}{P_{TX}}} \left(\int_0^\infty e^{\left[-\left(\frac{u}{P_{TX}}+\frac{1}{P_I}\right)v\right]} v\, dv + \eta^2 \int_0^\infty e^{\left[-\left(\frac{u}{P_{TX}}+\frac{1}{P_I}\right)v\right]} dv \right) \tag{20}$$

It is easy to check that the following two integrals have closed form solutions as:

$$\int_0^\infty e^{\left[-\left(\frac{u}{P_{TX}}+\frac{1}{P_I}\right)v\right]} dv = \left(\frac{u}{P_{TX}}+\frac{1}{P_I}\right)^{-1} = \frac{P_{TX}P_I}{P_I u + P_{TX}} \tag{21}$$

$$\int_0^\infty e^{\left[-\left(\frac{u}{P_{TX}}+\frac{1}{P_I}\right)v\right]} v\, dv = \left(\frac{u}{P_{TX}}+\frac{1}{P_I}\right)^{-2} = \frac{P_{TX}^2 P_I^2}{(P_I u + P_{TX})^2} \tag{22}$$

Finally the distribution of SINR is obtained as follows:

$$f_U(u) = \frac{1}{P_{TX}P_I} e^{-\frac{\eta^2 u}{P_{TX}}} \left(\frac{P_{TX}^2 P_I^2}{(P_I u + P_{TX})^2} + \frac{P_I P_{TX} \eta^2}{P_I u + P_{TX}} \right) \tag{23}$$

$$f_U(u) = e^{-\frac{\eta^2 u}{P_{TX}}} \left(\frac{P_{TX} P_I}{(P_I u + P_{TX})^2} + \frac{\eta^2}{P_I u + P_{TX}} \right) \tag{24}$$

This distribution has also been derived by Naghibi and Gross [3].

B. *Effective Capacity in uncorrelated Case using Laplace's Method*

In this section we first obtain the exact mean of the service rate. Then we compute the second moment of the service rate.

The mean of the service rate generally is given by:

$$E[s[i]] = \frac{1}{\ln 2} \int_0^\infty \left[\frac{A}{bu + c} + \frac{B}{(bu + c)^2} \right] \log_2(u + 1) e^{-au} du \tag{25}$$

where we have:

$$A = \eta^2$$
$$B = P_I P_{TX}$$
$$b = P_I$$
$$c = P_{TX}$$

We use the integration by parts to decompose the integrals in Eq. 25 into the sub-integrals as:

$$= \frac{A e^{\frac{ac}{b}}}{b \ln 2} \left[\text{Ei}(-au - a) \ln(u + 1) - \int_0^\infty \frac{\text{Ei}(-au - a)}{u + 1} du \right]$$

$$+ \frac{B}{\ln 2} \left[\frac{-a e^{\frac{ac}{b}}}{b^2} \text{Ei}(-au - a) \ln(u + 1) - \frac{1}{b} \frac{e^{-au} \ln(u + 1)}{bu + c} \right.$$

$$\left. + \frac{a e^{\frac{ac}{b}}}{b^2} \int_0^\infty \frac{\text{Ei}\left(-au - \frac{ac}{b}\right)}{u + 1} du + \frac{1}{b} \int_0^\infty \frac{e^{-au}}{(bu + c)(u + 1)} du \right] \tag{26}$$

After some simplifications we have:

$$E[s[i]] = \frac{1}{\ln 2} \left(\frac{Ba}{b^2} - \frac{A}{b} \right) e^{\frac{ac}{b}} \int_0^\infty \frac{\text{Ei}\left(-au - \frac{ac}{b}\right)}{u + 1} du$$

$$+ \frac{B}{b \ln 2} \int_0^\infty \frac{e^{-au}}{(bu + c)(u + 1)} du \tag{27}$$

We know that $\frac{Ba}{b^2} - \frac{A}{b} = 0$, hence:

$$E\left[s\left[i\right]\right] = \frac{B}{b\ln 2}\int_0^\infty \frac{e^{-au}}{(bu+c)(u+1)}du = \frac{B}{b\ln 2}I_4 \qquad (28)$$

$$\frac{A}{bu+c} + \frac{B}{u+1} = \frac{1}{(bu+c)(u+1)}$$

$$\rightarrow Au + A + Bbu + Bc = 1 \qquad (29)$$

$$A + Bb = 0, A + bc = 1 \rightarrow A = \frac{-b}{c-b}, B = \frac{1}{c-b} \qquad (30)$$

$$I_4 = \int_0^\infty \frac{e^{-au}}{(bu+c)(u+1)}du = \frac{1}{c-b}\left[\int_0^\infty \frac{-be^{-au}}{bu+c}du - \int_0^\infty \frac{e^{-au}}{u+1}du\right] \qquad (31)$$

$$I_4 = \frac{1}{c-b}\left[-e^{\frac{ac}{b}}E_1\left(\frac{ac}{b}\right) + e^a E_1(a)\right] \qquad (32)$$

Knowing that $\frac{B}{b} = P_{TX}$, the exact mean of the service rate in general values of signal-of-interest and interfering power is finally represented by:

$$E\left[s\left[i\right]\right] = \frac{P_{TX}}{(\ln 2)(P_{TX} - P_I)}\left[E_1\left(\frac{\eta^2}{P_{TX}}\right) - E_1\left(\frac{\eta^2}{P_I}\right)\right] \qquad (33)$$

Now we consider the variance in general case, to calculate the variance we first compute the second moment of the service rate using the following:

$$E\left[s^2\left[i\right]\right] = \int_0^\infty \left[\frac{A}{(bu+c)} + \frac{B}{(bu+c)^2}\right]\log_2^2(u+1)e^{-au}du \qquad (34)$$

Rewriting the $\log_2(u+1)$ in terms of $\ln(u+1)$ we have:

$$E\left[s^2\left[i\right]\right] = \frac{1}{\ln^2 2}\int_0^\infty \frac{A}{(bu+c)}\ln^2(u+1)e^{-au}du$$

$$+ \frac{1}{\ln^2 2}\int_0^\infty \frac{B}{(bu+c)^2}\ln^2(u+1)e^{-au}du \qquad (35)$$

Using integration by parts, the second moment becomes:

$$= \frac{A}{\ln^2 2}\left[\frac{e^{\frac{ac}{b}}}{b}\text{Ei}\left(-au-\frac{ac}{b}\right)\ln^2(u+1) - \frac{2e^{\frac{ac}{b}}}{b}\int_0^\infty \frac{\text{Ei}\left(-au-\frac{ac}{b}\right)\ln(u+1)}{u+1}du\right]$$

$$\frac{B}{\ln^2 2}\left[-a\frac{e^{\frac{ac}{b}}}{b^2}\text{Ei}\left(-au-\frac{ac}{b}\right)\ln^2(u+1) - \frac{e^{-au}\ln^2(u+1)}{b(bu+c)}\right]$$

$$+\frac{2ae^{\frac{ac}{b}}}{b^2}\int_0^\infty\frac{\text{Ei}\left(-au-\frac{ac}{b}\right)\ln\left(u+1\right)}{u+1}du+\frac{2}{b}\int_0^\infty\frac{e^{-au}\ln\left(u+1\right)}{\left(bu+c\right)\left(u+1\right)}du\Bigg] \tag{35}$$

$$\text{E}\left[s^2\left[i\right]\right]=\frac{2}{\ln^2 2}\left(\frac{Ba}{b^2}-\frac{A}{b}\right)e^{\frac{ac}{b}}\int_0^\infty\frac{\text{Ei}\left(-au-\frac{ac}{b}\right)\ln\left(u+1\right)}{u+1}du$$

$$+\frac{2B}{b\ln^2 2}\int_0^\infty\frac{e^{-au}\ln\left(u+1\right)}{\left(bu+c\right)\left(u+1\right)}du \tag{36}$$

We know that $\frac{Ba}{b^2}-\frac{A}{b}=0$. So we obtain.

$$\text{E}\left[s^2\left[i\right]\right]=\frac{2B}{b\ln^2 2}\int_0^\infty\frac{e^{-au}\ln\left(u+1\right)}{\left(bu+c\right)\left(u+1\right)}du \tag{37}$$

Laplace's Method [23] is one of the famous methods to approximate the computation of the integrals of the form $I\left(\lambda\right)=\int_a^b e^{-\lambda g\left(u\right)}f\left(u\right)du$ where

1) $g\left(u\right)$ is a smooth function and has a local minimum at u_c in the interval $[a, b]$. It means $g'\left(u_c\right)=0$ and $g''\left(u_c\right)>0$.
2) $f\left(u\right)$ is also smooth and $f\left(u_c\right)\neq 0$.

Laplace's method works in some steps that described below.
- We first obtain $g\left(u\right)$
- The minimum of $g\left(u\right)$ is computed.
- The second derivative of $g\left(u\right)$ at point u_c or $g''\left(u_c\right)$ is computed.
- $f\left(u_c\right)$ and $g\left(u_c\right)$ are obtained.

- Finally using the $I\left(\lambda\right)\approx e^{-\lambda g\left(u_c\right)}f\left(u_c\right)\sqrt{\frac{2\pi}{\lambda g''\left(uc\right)}}$ the integral is approximated.

We know that the second moment is given in Eq.37. Using the aforementioned procedure we have:

$$\text{E}\left[s^2\left[i\right]\right]=\frac{2B}{b\ln^2 2}\int_0^\infty\frac{\ln\left(u+1\right)e^{-au}}{\left(bu+c\right)\left(u+1\right)}du$$

$$=\frac{2B}{b^2\ln^2 2}\int_0^\infty\frac{\ln\left(u+1\right)e^{-au}}{\left(u+\frac{c}{b}\right)\left(u+1\right)}du \tag{38}$$

$$=\frac{2B}{b^2\ln^2 2}\int_0^\infty\frac{e^{-au}\ln\left(u+1\right)\left(u+1\right)^{\frac{kac}{b}}}{\left(u+\frac{c}{b}\right)\left(u+1\right)^{1+\frac{kac}{b}}}du=\frac{2B}{b^2\ln^2 2}I\left(a\right) \tag{39}$$

where $I\left(a\right)$ is:

$$I\left(a\right)=\int_0^\infty\frac{e^{-au}\ln\left(u+1\right)\left(u+1\right)^{\frac{kac}{b}}}{\left(u+\frac{c}{b}\right)\left(u+1\right)^{1+\frac{kac}{b}}}du \tag{40}$$

Using the Laplace's method in the first step we find the proper function $g\left(x\right)$.

$$=\frac{2B}{b^2\ln^2 2}\int_0^\infty\frac{e^{-au}e^{\ln\left(u+1\right)\frac{kac}{b}}\ln\left(u+1\right)}{\left(u+\frac{c}{b}\right)\left(u+1\right)^{1+\frac{kac}{b}}}du \tag{41}$$

$$=\frac{2B}{b^2\ln^2 2}\int_0^\infty\frac{\left(e^{-a\left(u-\frac{kc}{b}\ln\left(u+1\right)\right)}\right)\ln\left(u+1\right)}{\left(u+\frac{c}{b}\right)\left(u+1\right)^{1+\frac{kac}{b}}}du \tag{42}$$

Considering the Laplace's method, the $f(u)$ and $g(u)$ are:

$$g(u) = u - \frac{kc}{b} \ln(u+1) \tag{43}$$

$$f(u) = \frac{\ln(u+1)}{\left(u + \frac{c}{b}\right)(u+1)^{1 + \frac{kac}{b}}} \tag{44}$$

We now obtain the minimum of $g(u)$, we have:

$$g'(u) = 1 - \frac{kc}{b(u+1)} \tag{45}$$

$$g'(u) = 0 \rightarrow u_c = \frac{kc}{b} - 1 \tag{46}$$

$$g''(u) = \frac{kc}{b} \frac{1}{(u+1)^2} > 0 \rightarrow \tag{47}$$

$$g''(u_c) = \frac{b}{kc} \tag{48}$$

We then calculate the value of the $f(u_c)$ and also $g(u_c)$ at point u_c:

$$g(u_c) = \left(\frac{kc}{b} - 1\right) - \frac{kc}{b} \ln\left(\frac{kc}{b}\right) \tag{49}$$

$$f(u_c) = \frac{\ln\left(\frac{kc}{b}\right)}{\left(\frac{c}{b}(k+1) - 1\right)\left(\frac{kc}{b}\right)^{\frac{kc}{b}+1}} \tag{50}$$

Finally according to Laplace's method, $I(a)$ is approximated as:

$$I(a) \approx e^{-a\left[\left(\frac{kc}{b}-1\right) - \frac{kc}{b}\ln\left(\frac{kc}{b}\right)\right]} \frac{\ln\left(\frac{kc}{b}\right)}{\left(\frac{c}{b}(k+1)-1\right)\left(\frac{kc}{b}\right)^{\frac{kc}{b}+1}} \sqrt{\frac{2\pi kc}{ab}} \tag{51}$$

$$I(a) \approx e^a e^{\left(\frac{kac}{b}\left(\ln\left(\frac{kc}{b}\right)-1\right)\right)} \frac{\ln\left(\frac{kc}{b}\right)}{\left(\frac{c}{b}(k+1)-1\right)\left(\frac{kc}{b}\right)^{\frac{kc}{b}+1}} \sqrt{\frac{2\pi kc}{ab}} \tag{52}$$

And the second monent of the service rate becomes:

$$E\left[s^2[i]\right] = \frac{2P_{TX}}{\ln^2 2P_I} I(a) \approx \frac{2P_{TX}^2}{\ln^2 2P_I} e^{\frac{\eta^2}{P_{TX}}} e^{\left(\frac{k\eta^2}{P_I}\left(\ln\left(\frac{kP_{TX}}{P_I}\right)-1\right)\right)}$$

$$\frac{\ln\left(\frac{kP_{TX}}{P_I}\right)}{\left(\frac{P_{TX}}{P_I}(k+1)-1\right)\left(\frac{kP_{TX}}{P_I}\right)^{\frac{kP_{TX}}{P_I}+1}} \sqrt{\frac{2\pi k}{\eta^2 P_I}} \tag{53}$$

So the variance of the service rate based on Laplace's method is

$$\text{Var}\left[s[i]\right] = \frac{2P_{TX}^2}{\ln^2 2P_I} e^{\frac{\eta^2}{P_{TX}}} e^{\left(\frac{k\eta^2}{P_I}\left(\ln\left(\frac{kP_{TX}}{P_I}\right)-1\right)\right)}$$

$$\times \frac{\ln\left(\frac{kP_{TX}}{P_I}\right)}{\left(\frac{P_{TX}}{P_I}(k+1)-1\right)\left(\frac{kP_{TX}}{P_I}\right)^{\frac{kP_{TX}}{P_I}+1}} \sqrt{\frac{2\pi k}{\eta^2 P_I}}$$

$$- \left(\frac{P_{TX}}{(\ln 2)(P_{TX} - P_I)}\left[E_1\left(\frac{\eta^2}{P_{TX}}\right) - E_1\left(\frac{\eta^2}{P_I}\right)\right]\right)^2 \tag{54}$$

Using the mean and variance of the service rate computed in Eq.33 and Eq.54 respectively, the EC is easily obtained using $\alpha s (u) = E[s [i]] - \text{Var} [s [i]] \frac{u}{2}$

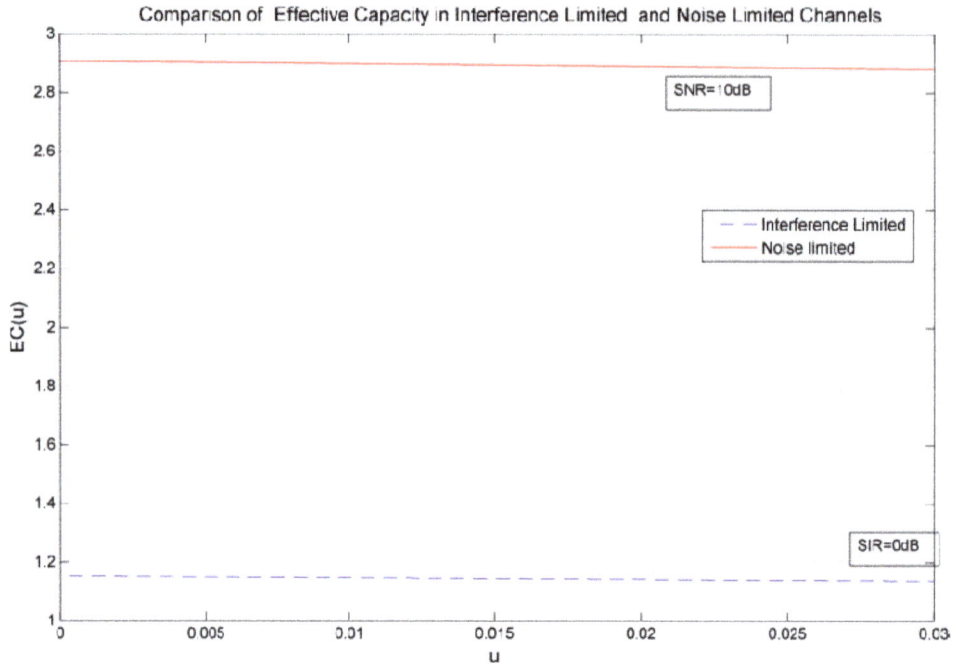

Figure 1. Validation of EC in Interference Limited uncorrelated Case

IV.   SOME EMPIRICAL RESULTS

In this part of the paper, we explain a procedure to assess the accuracy of the analytical EC model. We then evaluate the analytical delay probability and also the maximum sustainable rate that the wireless channel can support to meet the statistical delay probability.

A. Validation of the Analytical Model for the Effective Capacity

In this section we first provide a procedure to validate the analytical approach to EC that was described in the previous sections.

To validate the EC in uncorrelated case, we should generate the distribution of signal-of-interest and interfering signal independently. Since the instantaneous power in Rayleigh fading is exponentially distributed, we use the MATLAB build-in function to generate the exponential random variables.

The procedure to validate the EC in uncorrelated case is summarized as follows:

- First generate two independent exponential random variables. These exponential random variables represent signal-of-interest and interfering signals.
- The synthetic SINR is then simply computed as $\gamma_{\text{SINR}} [i] = \frac{P_{TX}}{P_I + \eta 2}$.
- Obtain the instantaneous service rate $s [i] = \log_2 (\gamma_{\text{SINR}} [i] + 1)$ is then obtained.
- Measure the sample mean and sample variance of service rate.
- Using sample mean and variance, estimate the empirical EC and compare it with the analytical EC model that we proposed in this paper.

The reader should note that in this paper we deal with uncorrelated Rayleigh fading. So the signal-of-interest and interfering signal are exponentially distributed and we can generate them in a straightforward manner.

Using the aforementioned procedure we validate the analytical approach that we proposed for the EC.

Figure 1 depicts the the EC in interference limited (IL) channels with $[\gamma_{SNR}]_{dB} = 0dB$ and noise limited (NL) channels with $[\gamma_{SNR}]_{dB} = 10dB$. As we see, EC in the IL channel is less than EC in the NL channel. The EC curves is both IL and NL channels are monotonically decreasing function of u that is expressed in terms of 1/kbits. The point in x axis that the EC is related to the QoS exponent. The QoS exponent is the inverse function of EC. It implies that the QoS exponent in IL channels is also less than NL channels. Since QoS exponent shows the exponential decay in delay probability, so the delay probability in IL channel will be higher than that of NL channels.

Figure 2 depicts the validation of EC based on Laplace's method. In this figure the dashed lines curves are EC obtained via simulation and the bold line curves are EC obtained via analytical approach. We believe that the analytical EC based on Laplace's method is a promising solution in more complicated fading scenarios that we have some kind of exponential function in the distribution of SINR.

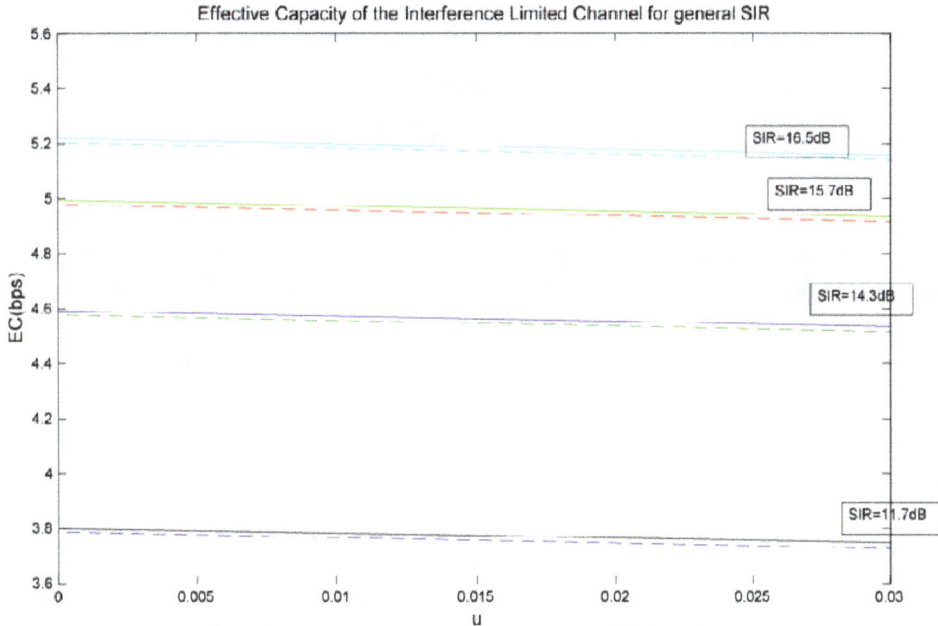

Figure 2. Validation of EC in uncorrelated case with general SIRs using Laplace's Method

B. Delay Probability of Interference Limited Channels in uncorrelated Case

Delay probability and MSR are the two performance metrics that use EC. If we know the EC function, arrival rate and target delay, using the results of large deviation principle the tail probability of delay in steady state can be obtained. From section II we know that the tail of delay probability is given by:

$$\Pr(D > D_t) \approx e^{-D_t \lambda \frac{2(E[s[t]] - \lambda)}{Var[[s_i]]}} \tag{55}$$

The steps for evaluating the tail probability of delay are as follows:

- First obtain mean of the service rate using:

$$E[s[i]] = \frac{P_{TX}}{(\ln 2)(P_{TX} - P_I)}\left[E_1\left(\frac{\eta^2}{P_{TX}}\right) - E_1\left(\frac{\eta^2}{P_I}\right)\right] \tag{56}$$

- Variance of the service rate is then computed. From section III we have:

$$\text{Var}[s[i]] = \frac{2P_{TX}^2}{\ln^2 2P_I}e^{\frac{\eta^2}{P_{TX}}}e^{\left(\frac{k\eta^2}{P_I}\left(\ln\left(\frac{kP_{TX}}{P_I}\right)-1\right)\right)}$$

$$\times \frac{\ln\left(\frac{kP_{TX}}{P_I}\right)}{\left(\frac{P_{TX}}{P_I}(k+1)-1\right)\left(\frac{kP_{TX}}{P_I}\right)^{\frac{kP_{TX}}{P_I}+1}}\sqrt{\frac{2\pi k}{\eta^2 P_I}}$$

$$-\left(\frac{P_{TX}}{(\ln 2)(P_{TX}-P_I)}\left[E_1\left(\frac{\eta^2}{P_{TX}}\right) - E_1\left(\frac{\eta^2}{P_I}\right)\right]\right)^2 \tag{57}$$

- Specify the arrival rate λ and target delay D_t.
- Finally use Eq.55 to calculate the analytical tail of delay.

Figure 3 represents the delay probability in the IL channels. According to this figure, the delay probability decreases as the average SIR (Signal to Interference Ratio) increases. From $[\gamma_{SIR}]_{dB}$ = 11.5dB to $[\gamma_{SIR}]_{dB}$ = 12.5dB the delay probability is reduced from 0.36 to 0.3. But if we increase SIR from $[\gamma_{SIR}]_{dB}$ = 15dB to $[\gamma_{SIR}]_{dB}$ = 16dB the delay probability will decrease from 0.263 to 0.256.

Figure 4 compares the delay probability in NL and NL channels. The x axis is the SIR in terms of dB for interference limited channels. The arrival rate is λ = 0.29 and target delay is D_t = 2ms. The red curve depicts the delay probability for the NL channel. NL channels here means discarding the interfering signal. For red curve

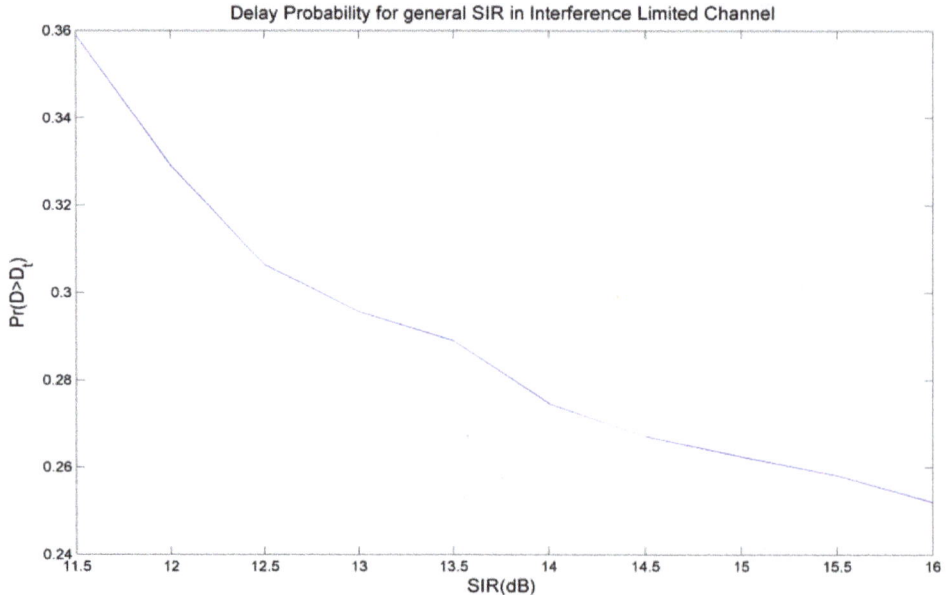

Figure 3. Delay Probability for general SIR in Interference Limited Channel Figure

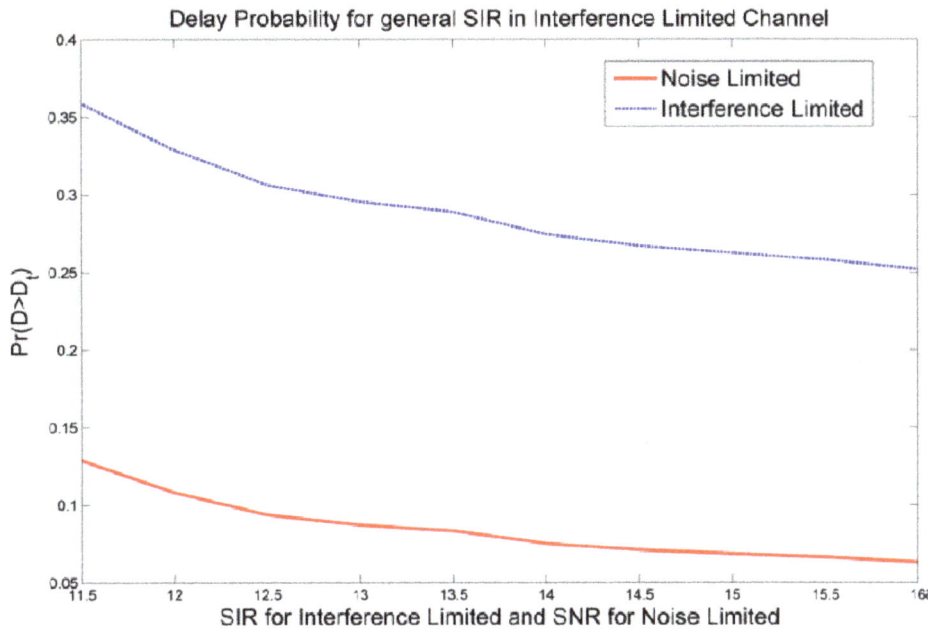

Figure 4. Comparison of Delay Probability in Interference Limited and Noise limited Channels

x axis is $[\gamma_{SNR}]_{dB}$ - 10dB. The delay probability for IL channels is generally higher than that of NL channels. But we will see in the next two figures the trade-off between target delay, arrival rate and delay probability.

Figure 5 illustrates the impact of different arrival rates on delay probability. The x axis is the SIR in terms of dB for IL channels. The target delay is $D_t = 8ms$. It is the same in all four delay probability curves. The red curves is the delay probability for the NL channel with $\lambda = 0.29$. Noise limited channels here means discarding interfering signal. For red curve x axis is γ_{SNR} - 10 in terms of dB. As it clear from the figure, the more arrival rate, the more tail probability of the delay we have. That is, in order to guarantee the same statistical delay in the presence of interference, we should decrease the transmission rate in the wireless networks.

Figure 6 investigates the impact of different target delays on delay probability. The x axis is the SIR in terms of dB for IL channels. The arrival rate is $\lambda = 0.1$. It is the same in all four delay probability curves. The red curves depicts the delay probability for the noise limited channel with $D_t = 6ms$. Noise limited channels here means discarding the interfering signal. For red curve x axis is γ_{SNR} - 10 in terms of dB. As it obvious from the figure, the more target delay, the more tail probability of the delay we have. That is, in order to guarantee the same statistical delay in the presence of interference, we should decrease the target delay. So in IL channels to keep the same level of statistical QoS we should tolerate the target delay or transmission rate.

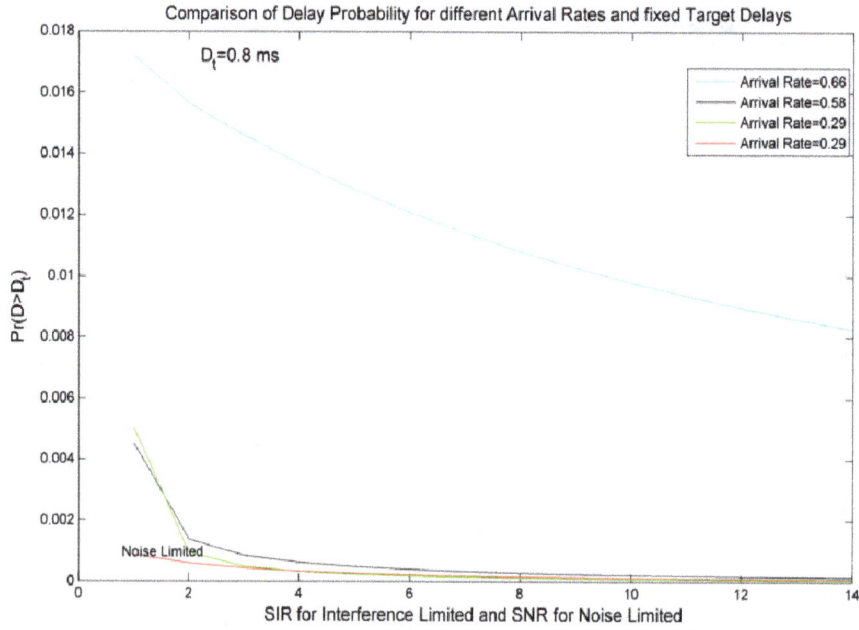

Figure 5. Comparison of Delay Probability for different Arrival Rates and a Fixed Target Delay

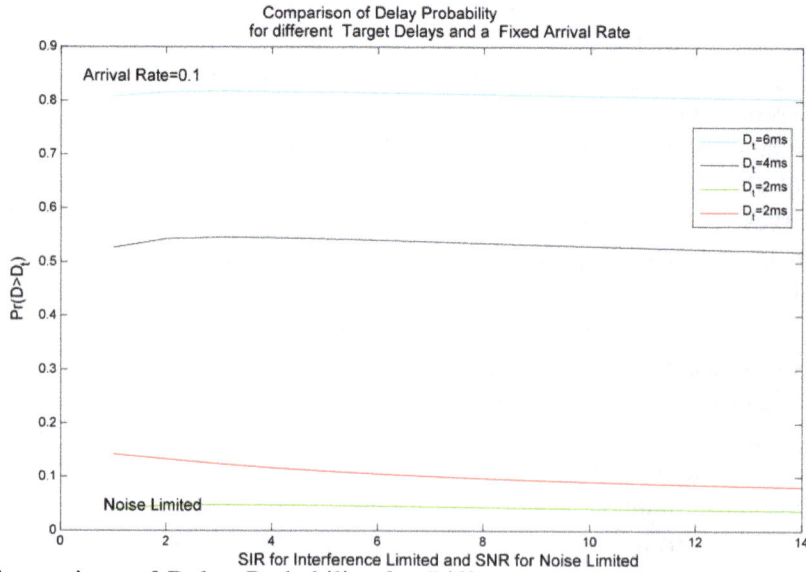

Figure 6. Comparison of Delay Probability for Different Target Delays and a Fixed Arrival Rate

C. Analysis of Maximum Sustainable Rate in uncorrelated Rayleigh Fading

MSR is defined as a maximum arrival rate that the wireless channel can support to fulfil the statistical QoS requirements [2]:

$$\lambda_{MSR} = \frac{\mathrm{E}\left[s\left[i\right]\right]}{2} + \frac{1}{2}\sqrt{\mathrm{E}\left[s\left[i\right]\right]^2 + 2\mathrm{Var}\left[s\left[i\right]\right]\left(\ln \epsilon / D_t\right)} \tag{58}$$

where E[s [i]] and Var [s [i]] are mean and variance of the service rate respectively. D_t represents the target delay, while $\epsilon = \Pr (D > D_t)$ is delay probability(or delay violation probability). It is evident from the Eq.58 that if $\epsilon \rightarrow 0$ or D_t is large then $\lambda_{M S R} \rightarrow$ E[s [i]]. In other words if we have loose QoS requirements, then EC approaches to mean of service rate.

If D_t or ϵ becomes lower, the wireless channel allows lower arrival rates in order to guarantee the QoS requirements and the $\lambda_{M S R}$ will be definitely less than E[s [i]]. It implies that if we have stringent QoS requirements, the EC will be between zero and mean service rate. Note that this conclusion is not valid for the family of EC curves that the normality assumption is not considered in their derivation. MSR can be a quantitative metric for wireless system designers and engineers to trade off between transmission rate and statistical QoS requirements in various wireless systems.

We compute the MSR using a procedure that is explained below:
- First obtain mean of the service rate using Eq.33.
- Variance of the service rate is then computed.
- Specify delay probability ϵ and target delay D_t.
- Finally use Eq.58 to calculate the analytical MSR.

In the above procedure, note that in the case we have only a single statistical QoS requirement.

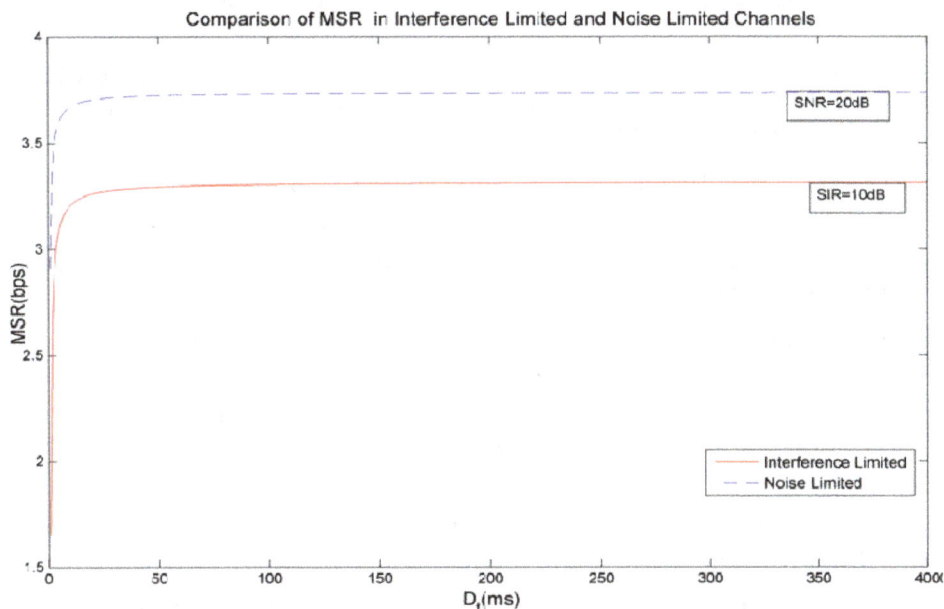

Figure 7. Comparison of MSR in Interference Limited and Noise Limited Channels

Figure 7 compares the MSR in IL channels with $[\gamma_{SIR}]_{dB}$ = 10dB and NL channels with $[\gamma_{SNR}]_{dB}$ = 20dB. From figure 1, we know that generally the EC curve for IL channels falls below the NL channels. The definition of MSR tells us that MSR is the maximum allowable rate that can be supported by the wireless channel to meet the statistical delay guarantee. This figures is in compatible with the EC curve in figure 1.

V. CONCLUSION

We investigated the statistical QoS provisioning in interference limited wireless networks. Specifically we evaluated the delay probability in block Rayleigh fading channels. To this end, we obtained the EC in uncorrelated Rayleigh faded signal of-interest and interfering signal. Using central limit theorem to obtain the EC, only the mean and variance of service rate should be computed. We computed the exact mean of the service rate. But obtaining the variance of the service rate was not a trivial problem. The deviation of the variance using Laplace's method was proposed in this paper. The analytical EC results were fitted to EC results obtained via simulation with a very good accuracy. Then an analysis of delay probability and MSR was proposed. We believe that the proposed analytical approach will open the path for the statistical QoS provisioning in interference limited wireless networks. Some of the future works are as follows:

- In this paper, we derived the EC in the IL channels in the presence of a single interferer. It is of interest to obtain the EC in the presence of multiple interferers.
- It is also of interest to invest more on Laplace's method and apply this method for derivation of the EC for other fading scenarios such as Nakagami-m fading.

REFERENCES

[1] D. Wu, and R. Negi, "Effective Capacity: A Wireless Link Model for Support of Quality of Service", *IEEE Transactions on Wireless Communications*, Vol. 2, No. 4, 2003.

[2] B. Soret, C. Aguayo-Torres and T. Entrambasaguas, "Capacity with Explicit Delay Guarantees for Generic Sources over Correlated Rayleigh Channel", *IEEE Transactions on Wireless Communications,* Vol. 9, No. 6, 2010.

[3] F. Naghibi and J. Gross, "How bad is Interference in IEEE 802.16e systems?" , *In Proc. of European Wireless Conference*, 2010.

[4] D. Wu and R. Negi , "Effective Capacity-Based Quality of Service Measures for Wireless Networks ", *Mobile Networks and Applications* , Vol. 11, No. 1, 2006.

[5] Q. Wang, D. Wu and P. Fan, "Effective Capacity of a Correlated Rayleigh Fading Channel", *Wireless Communications and Mobile Computing* , 2010.

[6] J. Tang and X. Zhang, "Cross-Layer Resource Allocation over Wireless Relay Networks for Quality of Service Provisioning", *IEEE Journal of Selected Areas on Communications* , Vol. 25, No. 4, 2007.

[7] D. Tse and P. Wiswanath, "Fundamentals of Wireless Communication ", *Cambridge University Press*, 2005.

[8] D. P. Pezaros, M. Sifalakis and D. Hutchison ,"On the Long-Range Dependent Behaviour of Unidirectional Packet Delay of Wireless Traffic", *In Proc. of IEEE Globecom*, 2007.

[9] O. Cappe, E. Moulines, J. Pesquet, A. P. Petropulu and Y. Xueshi, "Long-range Dependence and Heavy-tail Modeling for Teletraffic Data", *IEEE Signal Processing Magazine*, Vol. 19, No. 3, 2002.

[10] N. Jindal, J. Andrews and S. Weber , "Optimizing the SINR Operating Point of Spatial Networks", *In Proc. of IEEE Information Theory and Applications Workshop*, 2007.

[11] M. Andrews, K. Kumaran, K. Ramanan, A. Stolyar, P. Whiting, and R. Vijayakumar, "Providing Quality of Service over a Shared Wireless Link", *IEEE Communications Magazine*, Vol. 39, No. 2, 2001.

[12] V. Bharghavan, L. Songwu and T. Nandagopal, "Fair Queuing in Wireless Networks: Issues and Approaches ", *IEEE Personal Communications*, Vol. 6, No. 1, 1999.

[13] S. Akin and M.C. Gursoy, "Effective Capacity Analysis of Cognitive Radio Channels for Quality of Service Provisioning ", *IEEE Transactions on Wireless Communications*, Vol. 9, No. 11, 2010.

[14] L. Lingjia and J. F. Chamberland, "On the Effective Capacities of Multiple-Antenna Gaussian Channels ", *In Proc. IEEE International Symposium on Information Theory*, 2008.

[15] A. Balasubramanian and S. L. Miller ,"The Effective Capacity of a Time Division Downlink Scheduling System", *IEEE Transactions on Communications*, Vol. 58, No. 1, 2010.

[16] Q. Du and X. Zhang ,"Resource Allocation for Downlink Statistical Multiuser QoS Provisioning in Cellular Wireless Networks ", *In Proc. of INFOCOM*, 2008.

[17] M. Xie and M. Haenggi, "Performance Analysis of a Priority Queueing System Over Rayleigh Fading Channels*", in Proc. Of Annual Allerton Conference on Communication, Control, and Computing, ,* 2003.

[18] C. Li, H. Che and S. Li, , "A Wireless Channel Capacity Model for Quality of Service", *IEEE Transactions Wireless Networks*, Vol.6, No.1, 2007.

[19] Y. Chen , J. Chen and Y. Yang , "Multi-Hop Delay Performance in Wireless Mesh Networks", *Mobile Networks and Applications*, Vol. 13, No. 1, 2008.

[20] L. Musavian and S. Aissa, "Effective Capacity of Delay-Constrained Cognitive Radio in Nakagami Fading Channels ", *IEEE Transactions on Wireless Communications*, Vol.9, No.3, 2010.

[21] E. A. Jorswieck, R. Mochaourab and M. Mittelbach ,"Effective Capacity Maximization in Multi-Antenna Channels with Covariance Feedback" , *IEEE Transactions on Wireless Communications*, Vol. 9, No. 10, 2010.

[22] M. K. Simon and M. S. Alouini , "Digital Communication over Fading Channels " , *J. Wiley and Sons*, 2th Edition, 2005.

[23] C. M. Bender and S. A. Orszag, "Advanced Mathematical Methods for Scientists and Engineers" , *Springer Verlag*, 1999.

[24] D. Wu and R. Negi ,"Utilizing Multiuser Diversity for Efficient Support of Quality of Service Over a Fading Channel ", *IEEE Transactions on Vehicular Technology*, Vol. 54, No. 3, 2005.

[25] M.C. Gursoy ,"MIMO Wireless Communications Under Statistical Queueing Constraints ", *IEEE Transactions on Information Theory*, Vol. 57, No. 9, 2011.

[26] B. Soret, M. A. Torres, and J. T. Entrambasaguas, "Analysis of the Trade-off between Delay and Source Rate in Multiuser Wireless Systems ", *EURASIP Journal on Wireless Communications and Networking*, 2010.

[27] R. Shaolei and K. B. Letaief, "Maximizing the Effective Capacity for Wireless Cooperative Relay Networks with QoS Guarantees " , *IEEE Transactionn on Communications*, Vol. 57 , No. 7, 2009.

[28] A. Dembo and O. Zeitouni , "Large Deviations Techniques and Applications ", *Springer Verlag*, 2th Edition, 2010.

[29] A. Popoulis and S.U. Pillai, " Probability Random Variables and Stochastic Processes ", *MC Graw Hill*, 2002.

7

EVALUATION AND ENHANCEMENT OF PERFORMANCE METRICS FOR IR AND BWR IN MAC 802.16

R. Bhakthavathsalam[1] and Khurram J.Mohammed[2]

[1]Supercomputer Education and Research Centre
Indian Institute of Science, Bangalore, India.
bhaktha@serc.iisc.ernet.in
[2]Ghousia College of Engineering, Ramanagaram, India.
khurramashrafi@gmail.com

ABSTRACT

In the IEEE 802.16 standard, Initial Ranging (IR) is defined as the mechanism of acquiring the correct timing offsets and power adjustments such that the Subscriber Station (SS) is co-located with the Base Station (BS). Bandwidth Request (BWR) is the process by which the communication is established between BS and SS requesting the uplink bandwidth allocation. In this, we evaluate the performance of these two schemes based on the metrics of delay and throughput. Then we enhance their performance by incorporating circularity. Circularity is a paradigm that allows the identification of specific groups of packets or events. Using this, we introduce delay control and backoff window control in the case of IR and selective dropping of BWR packets. This new paradigm reduces the collisions among request packets and thereby, improves both these mechanisms. The evaluation and enhancement are performed through extensive simulated studies.

KEYWORDS

IEEE 802.16, MAC sub layer, Contention Resolution, Initial Ranging, Bandwidth Request, Circularity, Network Simulator 2

1. INTRODUCTION

The Institute of Electrical and Electronics Engineers (IEEE) 802.16 standard for Wireless Metropolitan Area Networks specifies the most recent technical features of wireless technology. Originally intended for Fixed Broadband Wireless Access (FBWA) networks and as a wireless competitor for wire-line DSL and cable modem access in particular in rural and low-infrastructure areas, the most recent stage of the IEEE 802.16 standard also provides mobility support mainly intended for nomadic users or users with little mobility. Worldwide Interoperability for Microwave Access (WiMAX) is a group established to empower the interoperability and promote the commercialization of products based on theIEEE 802.16 standard. The current IEEE 802.16-2004 standard with the extensions for mobility support amended in the IEEE 802.16e-2005 standard is the basis for two classes of WiMAX certified products [1]. The Orthogonal Frequency Division Multiplexing (OFDM) part of IEEE 802.16-2004 is known as Fixed WiMAX and the Orthogonal Frequency Division Multiple Access (OFDMA) part of IEEE 802.16e-2005 is known as Mobile WiMAX [2], [3].

1.1.Overview of 802.16 PHY and MAC layers

The IEEE 802.16 standard specifies four physical layers namely Single Carrier (SC), Single Carrier-a (SCa), single carrier transmission in line-of-sight and non-line-of-sight environments. A common Medium Access Control (MAC) sub layer is defined for all physical layers with only small adaptations to the different physical layers. The standard stipulates two   types of operation point-to-multi-point (P2MP) and mesh mode. In the following we focus on point-to-multipoint communication and the SC, SCa and OFDM physical layers. In P2MP mode of operation, the connection is established between a base station and subscriber stations/base stations or both. The MAC sub layer is present within Data Link layer of Open Systems Interconnection (OSI) model. MAC sub layer has a great role in OSI model. It serves as a link between lower hardware oriented PHY layer and other upper software driven layers. IEEE 802.16 MAC sub layer is further divided into three parts as Service specific Convergent Sub layer (CS), Common Part Sub layer (CPS), and Privacy Sub layer (PS). The MAC layer functions are Transmission Scheduling, Admission Control, Link Initialization, Fragmentation, Retransmission and support for integrated Voice/Data connections [4].

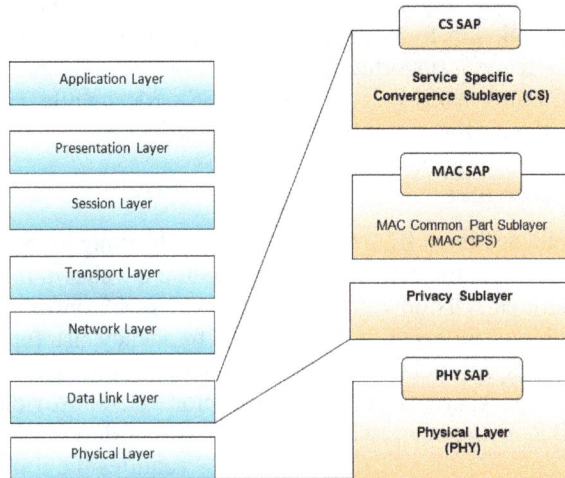

Figure 1. Protocol Layering

Figure 1 shows the positions of the different layers and sub layers mentioned above. Three types of MAC management connections are possible in MAC 802.16 to exchange control messages between the BS and SS. These connections specify different Quality of Service (QoS) requirements, needed at different management levels [5]. The BS is identified by a 48-bit Base Station Identifier (ID) and each SS is identified by a 48-bit MAC address. The 16-bit Connection Identifier (CID) identifies each connection with BS and SS in a session. In P2MP type of operation, the connection between BS and SS is established based on Request/Grant mechanism. The frame structure shown in Figure 2 best explains the operation of MAC layer [6]. Data units between the BS and SSs are exchanged in units of fixed-length frames. The frame length consists of a downlink and an uplink sub frame.

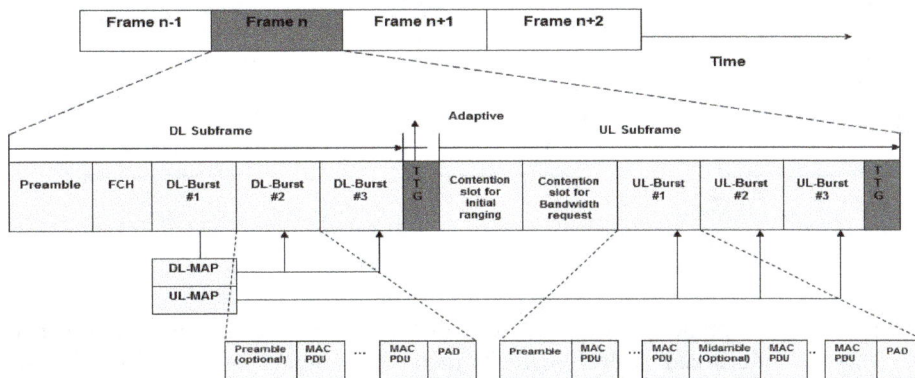

Figure 2. Frame Structure for IEEE 802.16

The transfer of data from BS to SSs takes place in the downlink sub frame and the uplink transfer occurs in opposite direction. The border between two sub frames is adaptive according to the data amount in downlink and uplink. An integer number of OFDM symbols carrying MAC datagrams are contained in each frame. These OFDM symbols comprise MAC resources available for data transmissions. Sharing of MAC resources between SSs is centralized in the BS. Downlink map (DL-MAP) and uplink map (UL-MAP) signaling messages are used to inform SSs about bandwidth allocations in downlink and uplink. At the beginning of the downlink sub frame a preamble for synchronization is transmitted, followed by the broadcast MAC signaling messages. Thereafter the bursts of user data are transmitted to the corresponding SS switch the use of different modulation types and coding rates. The Transmit/Receive (Tx/Rx) transition gaps between the sub frames allow the stations to switch between transmission and reception modes. The uplink sub frame begins with contention intervals scheduled for initial ranging and bandwidth request opportunities. IR during network entry accommodates the adjustments of time and power parameters by the new SSs. Meanwhile, BWR fulfills the needs of a SS for necessary resources for its transmission to the BS.

1.2.Network Entry Procedure

The SS needs to complete the network entry process for establishing its connections in the entire network. The different stages in the network entry procedure are shown in the Figure below [7]. The first stage of network entry is downlink (DL) channel synchronization. To begin with the SS scans for the available channels in the prescribed frequency list for its operations in the WiMAX network. The SS synchronizes at PHY level with the use of periodic frame preamble after identifying a DL channel. The DL channel descriptor (DCD) provides the information on modulation and other DL parameters, while the UL channel descriptor (UCD) contains the similar information for uplink. In Initial Ranging, the SS acquires the timing offsets and power adjustments from the BS.

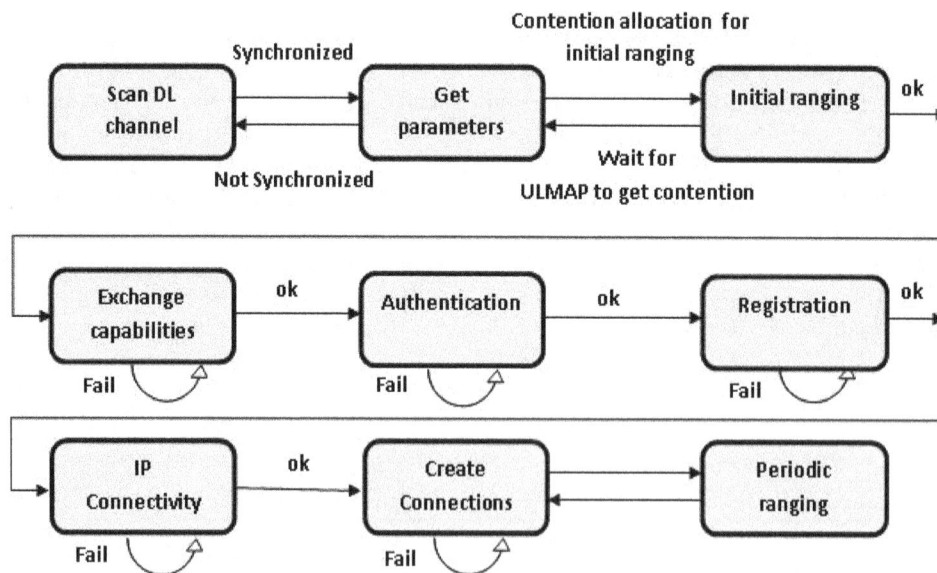

Figure 3. Network Entry Process

This enables the SS to properly communicate with the BS. IR is a very important part of the network entry procedure and is dealt with in more detail in the next section. The step of Exchanging Capabilities follows the IR completion. The Exchanging Capabilities involve rates and duplexing methods, modulation level, and coding scheme. Next follows Authentication wherein, the BS validates the SS, determines the ciphering algorithm to be used, and sends an authentication response to the SS. In the following Registration step, the SS exchanges a registration request message to the BS and the BS in return sends the registration response. The registration response message includes the secondary management CID of the SS. Using this, a SS is allowed entry into the network and the SS is said to be manageable. Next in Internet Protocol (IP) Connectivity, the SS gets the IP address via DHCP. The SS also fetches the additional operational parameters using TFTP. After completing the IP connectivity step, transport connections are formed in the final Connection Creation.

2. IR AND BWR MECHANISMS

Initial Ranging is an important part of the Network Entry procedure performed by the SSs, upon power up, in IEEE 802.16 networks. In the IR procedure, the correct timing offsets and power adjustments are obtained from the BS so that the SS can successfully transmit data to the BS. It occurs after the SS has synchronized with a downlink channel from the BS and has obtained the uplink transmit parameters from the UCD MAC management message. After this the SS will scan the UL-MAP message to find an Initial Ranging Interval, consisting of one or more transmission opportunities allocated by the BS. Then SS initiates the IR procedure and assembles a Ranging Request (RNG-REQ) message and sends to the BS.

The SS sends this message as if it is co-located with the BS. This is done by setting the initial timing offset to the internal fixed delay equivalent to co-locating the SS next to the BS. The SS calculates the maximum transmit signal strength for initial ranging and transmits the RNG-REQ

message at a power level below this measured at the antenna connector. In case a response is not received from the BS, the next RNG-REQ message is sent at the next higher power level in the next appropriate Initial Ranging Interval. In case it receives a response from the BS, depending the contents of the response the SS does the following. If the Ranging Response (RNG-RSP) message contains the frame number in which the RNG-REQ message was sent, the SS will consider the previous attempt to be unsuccessful. Nevertheless, it will make the adjustments specified in the RNG-RSP message. If the RNG-RSP message contains the MAC address of the SS, then the request attempt will be considered successful. When the RNG-REQ message is successfully received by the BS, it will send an RNG-RSP message using the Initial Ranging CID. At this point the BS starts using Invited Initial Ranging Intervals addressed to the Basic Connection Identifier of the SS to complete the process of IR. But if the status in the RNG-RSP is success, the IR procedure will be completed. On receiving an RNG-RSP message with continue status, the SS first makes the power level and timing adjustments. Then it retransmits another RNG-REQ using the Basic CID assigned to it. The BS yet again sends an RNG-RSP message containing additional fine tuning, if required. This exchange of RNG-REQ and RNG-RSP messages continues till an RNG-RSP message with status success is received by the SS or the BS aborts the IR procedure.

Whenever the SS has to transmit the request packets it performs the Truncated Binary Exponential Backoff procedure. This method is the contention resolution procedure used in IEEE 802.16 networks [8], [9]. The minimum backoff window and the maximum backoff window are both controlled by the BS. Initially the SS set sits backoff window to the minimum possible backoff window. Now the SS randomly selects a number from this backoff window. This number selected indicates the number of IR transmission opportunities that the SS must defer before transmitting the request packet. After the selected number of transmission opportunities is deferred, the SS transmits the RNG-REQ message.

After transmitting the request message, the SS waits for a response message from the BS. If the RNG-RSP message is received from the BS before the specified timeout then the contention resolution is considered to be a success. If not, the SS doubles its backoff window until the maximum backoff window is reached. It then randomly selects another number from this new window and the deferring process is repeated. This may happen due to the collision of RNG-REQ packets or due to the loss of RNG-RSP messages. There exists a maximum limit for the number of such IR retries allowed. If this limit is reached by an SS, then the particular downlink channel being used is marked as unusable and it begins scanning for a new downlink channel. The backoff windows always increase in terms of powers of two. For example, the backoff window at a particular instant for an SS is lying between 0 and 31 (0 to 2^5 - 1) and the random number picked is 11. The SS has to defer a total of 11 Initial Ranging Intervals before transmitting the RNG-REQ packet. This may require the SS to defer IR intervals over multiple frames. In case a collision is detected, the backoff window is increased twofold. Next a random number is picked between 0 and 63 and the deferring process is continued repeatedly for a maximum of 16 times after which the uplink channel is restarted from the beginning. These basic steps of the contention resolution procedure are identical to both Initial Ranging and Bandwidth Request. These steps are shown in Figure 4.

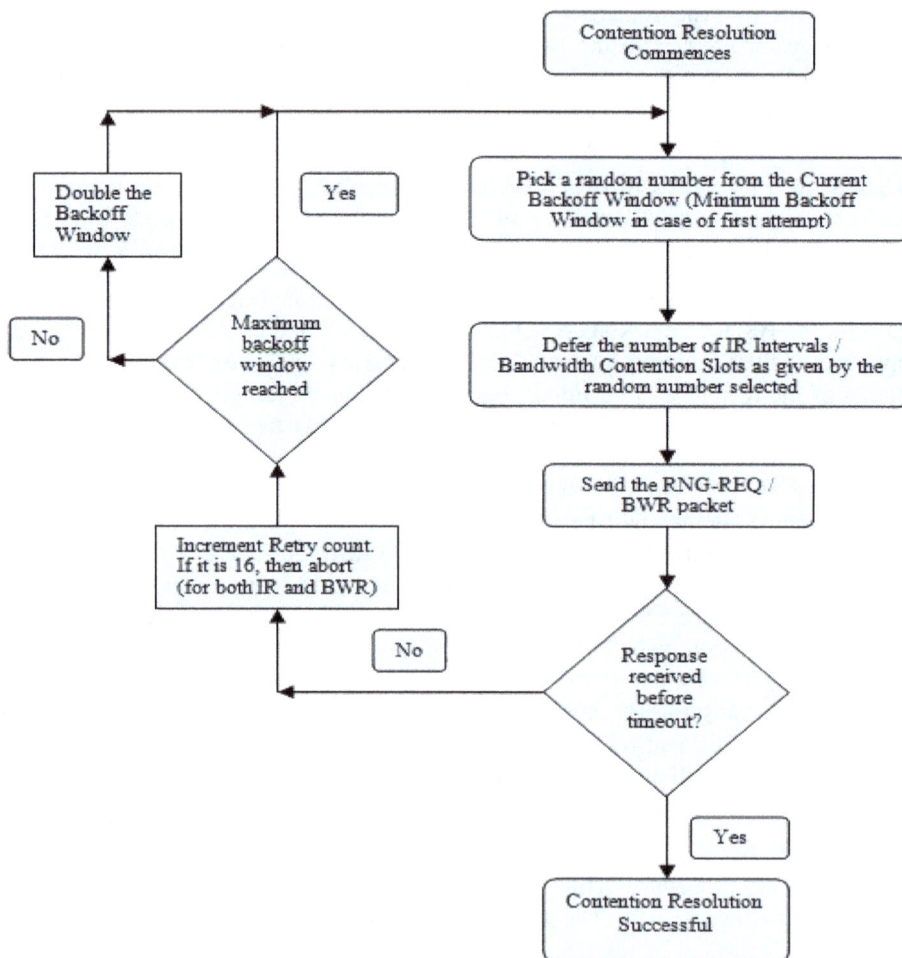

Figure 4. Contention Resolution Process

Bandwidth Request deals with the process through which the SS demands the bandwidth requirements for the uplink channel for data transmission [10], [11]. In Initial Ranging, the RNG-REQ comprises only single type of request, whereas BWR packets contain multiple types such as stand-alone, piggybacked BWR packets and also a either incremental or aggregate BWR messages. In all cases, the data requests are made in terms of the number of bytes required to be sent, but not in terms of the channel capacity. This is due to fact that the uplink burst profile keeps changing dynamically. In the uplink sub frame of the TDD frame structure of 802.16, there are a number of Contention slots present for the purposes of BWR [12].

In contention based BWR mechanism, an SS tries to transmit its BWR packet to the BS in one of the contention slots found in the uplink sub frame. Under contention free condition, if a BWR packet is successfully received by the BS, the requested bandwidth is fully available. Once the bandwidth is reserved for the SS, it can freely transmit its data without contention in the next frame. In case of multiple SSs sending their request messages to the same BS simultaneously, it gives rise to collisions. So, here again the contention is resolved by using the truncated binary exponential backoff procedure, which is the same as in the case of IR. The BS decides the range of backoff windows from minimum to maximum numbers.

Also, the number of bandwidth request trials is limited to 16. One can measure the performance of a network in many ways. This is mainly due to the variety of networks that are present. Some of the metrics that can be used to measure network performance are delay, throughput, availability, bandwidth, etc. We have selected delay and success-ratio of the Ranging Request packets as the metrics for measuring the performance of the Initial Ranging mechanism. These two metrics are also used to compare the performances of IR before and after incorporating the paradigm of circularity. Also, the delay metric is used for the BWR mechanism to show the effect of incorporating circularity.

3. ANALYSIS OF IR AND EVALUATION OF TOTAL DELAY

In this section we analyze the IR scheme and evaluate mathematically the delay involved in the procedure. First we structure the IR mechanism as a set of distinct states with transitions among these states. Then this information is used to model IR as a Markov process. In a Markov process, the probability of the system making a transition to a particular state depends only on the state the system is currently in [13]. Also, in this Markov process, we calculate the delays associated with the transitions between the states. Finally, by making use of the delays and probabilities associated with each of the transitions, mathematical equation describing the total IR delay is obtained.

3.1.Modeling IR as a Markov Process

Markov processes offer very flexible, powerful, and efficient tools for describing and analyzing the dynamic system properties. Thus they help us to easily estimate the performance and dependability measures. Markov processes unify the fundamental theory underlying the concept of queuing systems. The notation of queuing systems has been considered as a high-level specification technique for (a sub-class of) Markov processes. The queuing systems can be mathematically evaluated by modeling them as Markov process. But besides highlighting the computational relation between Markov processes and queuing systems, it is worthwhile pointing out also that fundamental properties of queuing systems are commonly proved in terms of the underlying Markov processes. This type of Markov processes is also possible even when queuing systems exhibit properties such as non-exponential distributions that cannot be represented directly by discrete-state processes. Markovian models, such as embedding techniques or supplementary variables, can be used in such cases. These models serve as the theoretical framework for determining the correctness of computational methods applied directly to the analysis and evaluation of queuing systems [14].

A stochastic process is defined as a family of random variables $\{X_t: t \, \varepsilon T\}$ where each random variable X_t is indexed by parameter t belonging to T, which is usually called the time parameter if T is a subset of R+ = [0, infinity). The set of all possible values of X_t (for each t εT) is known as the state space S of the stochastic process. Many numbers of stochastic processes belong to an important class of Markov models. Thus, a Markov model represents a stochastic process where the future state of the process depends only the current state and not on the history of the system. A Markov process is a memory-less stochastic process [13].In the reference [15], the Backoff procedure is modeled as a Markov process. Here, we extend this to incorporate the entire Initial Ranging mechanism. After analyzing IR, we enumerate the following states as well as transitions needed for modeling the procedure.

1. Waiting for UL-MAP. This is also the start state.
2. SS is performing Backoff procedure.
3. Waiting for an RNG-RSP message from BS.
4. Continue
5. Success State – Wait for CDMA Allocation IE.
6. Abort – Start network entry procedure at a different DL channel
7. Waits for RNG-RSP again.
8. Proceed to next phase of network entry
9. Commence Periodic Ranging

The transitions among the states are as follows. In State 1, the SS waits for a UL-MAP. After receiving this message it makes a transition to State 2. Transmission of CDMA code occurs at end of State 2. Also a timer is set for waiting for RNG-RSP message. This transition leaves the system in State 3. When in State 3, if the timer for RNG-RSP expires then SS increments the power level and goes back to State 1. When in State 3, if RNG-RSP is obtained with Ranging code as well as the Ranging slot, then it makes a transition to State 4. Here the necessary adjustments specified in RNG-RSP are made and system moves to State 1. When in State 3, if RNG-RSP is obtained with success status, then the system transits to State 5. Here it waits for CDMA Allocation IE. After reception it sends RNG-REQ message on the allocated bandwidth and moves to State 7.When in State 7, on reception of RNG-RSP with success status it moves to State 8. On reception of RNG-RSP with continue status it moves to State 9. Else on reception of RNG-RSP with abort status, it goes to State 6 and SS starts the network entry procedure again. When in State 3, if RNG-RSP is obtained with abort status then the system goes to State 6 and SS starts the network entry procedure again. The following matrix diagram shows the Markov process that represents the Initial Ranging procedure of IEEE 802.16 network standard. The Transition Probability Matrix corresponding to the states in this process is shown in table 1.

Table 1. Probability matrix corresponding to the states

	1	2	3	4	5	6	7	8	9
1	0	1	0	0	0	0	0	0	0
2	0	0	1	0	0	0	0	0	0
3	a1	0	0	a2	a3	a4	0	0	0
4	1	0	0	0	0	0	0	0	0
5	0	0	0	0	0	0	1	0	0
6	0	0	0	0	0	1	0	0	0
7	0	0	0	0	0	b3	0	b1	b2
8	0	0	0	0	0	0	0	1	0
9	0	0	0	0	0	0	0	0	1

The transitions out of states 3 and 7 are non-deterministic and therefore we have used algebraic symbols to represent the probabilities associated with outgoing transitions. Using these probabilities, we design the Markov process representation of IR as shown in Figure 5. The states 6, 8 and 9 lead out of the IR mechanism and are the absorbing states. Next,   we use the transition matrix obtained above to obtain the overall delay formula. For this, we first need to tabulate the delays involved in the individual states. The states 6, 8 and 9 are absorbing states. For these states, the probability of transition to themselves is 1.

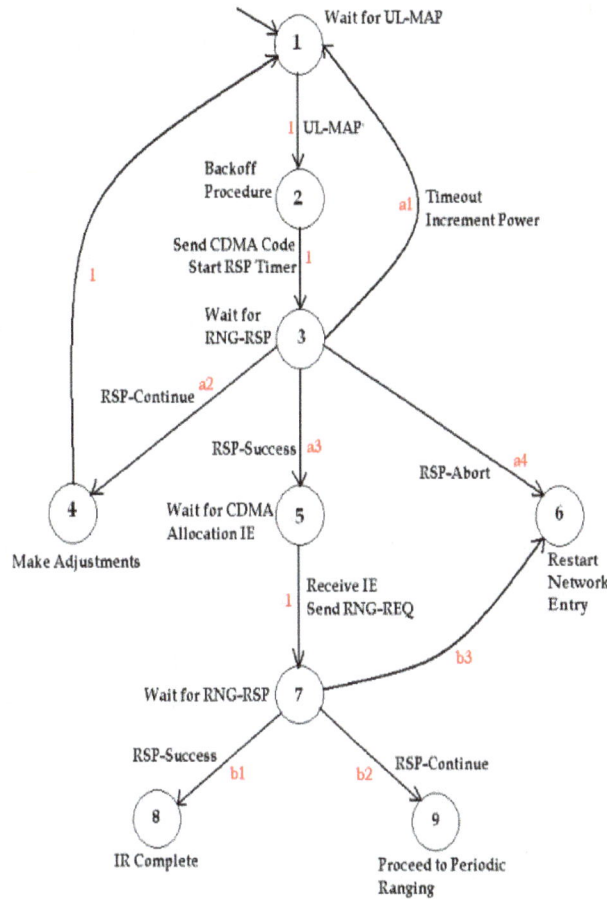

Figure 5. IR Procedure as a Markov Process

Table 2. Details of the delays involved along with the associated probabilities

Delay Involved	Probabilities
UL-MAP Reception (1 to 2)	1
Backoff Delay + Sending CDMA (2 to 3)	1
RNG-RSP Timeout (3 to 1)	a1
RNG-RSP Reception + Processing (3 to 4, 5 or 6)	a2,a3,a4
IE Allocation Delay + Sending RNG-REQ (5 to 7)	1
RNG-RSP Reception + Processing (7 to 8, 9 or 6)	b1,b2,b3

The details of the delays involved along with the associated probabilities are given in the table below. In states 3 and 7, the outgoing probabilities are marked with algebraic symbols a1 to a4 and b1 to b3. This is because the probabilities of the transitions originating from these states are non-deterministic in nature. Nevertheless, the sum of probabilities of all transitions originating from states 3 and 6 are still equal to 1.

The numerical values of the delays involved are expressed below where the symbol → indicates the transition from one state to other as depicted above in Figure 5 (i→ j, $1 \leq i, j \geq 9$):

1. UL-MAP Reception (UL_M) = 5ms (Maximum of one complete frame length)[1 → 2]
2. CDMA Sending Time = $CDMA_{ST}$ = Transmission Time = 5ms/2 = 2.5ms
   [Length of UL subframe= (Frame Length/2) where frame length is 5ms][2 → 3]
3. RNG-RSP Timeout = T3 = 200 ms[3 → 1]
4. RNG-RSP Reception (RSP) + Processing (average value)
   = T3/2 + Max. RNG-RSP Processing Time/2
   = 100 ms + 10ms/2 = 105 ms[3 → 4, 5, or 6]
5. CDMA Allocation IE delay = $CDMA_{IE}$ = 5ms (same as 1)[5 → 7]
6. Sending RNG-REQ (Same as 2) = 2.5ms [5 → 7]
7. RNG-RSP Reception + Processing (average value) = 105ms [7 → 8 or 9]

We assume that the delay involved for making changes at SS is negligible compared to the other delays involved.

3.2. Mathematical Derivation of the Backoff delay

Consider the first time an SS enters Backoff procedure. Let the initial Contention Window be w_0. The random number will be picked in the range [0, w_0-1]. Let this random number be called 'k'. The SS has to defer a total of k contention slots. Let the number of CS's in a frame be n_{cs}. The number of frames that have to be deferred is k/n_{cs}. The delay involved here will be (k/n_{cs})*frame length. After k/n_{cs} frames have passed the SS defers a further (k modulo n_{cs}) Contention Slots. The delay involved here is equal to (k mod n_{cs})*T_{cs}, where T_{cs} is the length of one CS.

$$\text{Total delay so far} = (k/ n_{cs})*\text{frame length} + (k \bmod n_{cs})* T_{cs} \tag{1}$$

Here the value of k can vary from 0 to w_0-1. Thus, we take an average of the delay (AD_0) over the random number k.

$$AD_0 = (1/w_0) * \sum_{k=0}^{(W0-1)} \left(\frac{k}{n_{cs}}\right) * \text{frame length} + (k \bmod n_{cs}) * T_{cs} \tag{2}$$

Next we make an assumption that the probability of a successful transmission in a CS is 'p'. Thus, probability of failure will be '1-p'. In case of a failure the contention window is doubled in size. Let the new window be [0, w_1-1]. Similar to equation (1a), the delay involved will be

$$AD_1 = (1/w_1) * \sum_{k=0}^{(W_1-1)} \left(\frac{k}{n_{cs}}\right) * \text{frame length} + (k \bmod n_{cs}) * T_{cs} \tag{3}$$

Here $w_1 = 2 * w_0$

Again there could be success or failure. So, it will enter the third Backoff window phase [0, w_2-1]. Continuing in this fashion, we get the following delays for the next three phases.

$$AD_2 = (1/w_2) * \sum_{k=0}^{(W_2-1)} \left(\frac{k}{n_{cs}}\right) * \text{frame length} + (k \bmod n_{cs}) * T_{cs} \tag{4}$$

Here $w_2 = 2 * w_1$

$$AD_3 = (1/w_3) * \sum_{k=0}^{(W_3-1)} \left(\frac{k}{n_{cs}}\right) * \text{frame length} + (k \bmod n_{cs}) * T_{cs} \tag{5}$$

Here $w_3 = 2 * w_2$

$$AD_4 = (1/w_4) * \sum_{k=0}^{(W_4-1)} \left(\frac{k}{n_{cs}}\right) * \text{frame length} + (k \bmod n_{cs}) * T_{cs} \tag{6}$$

Here $w_4 = 2 * w_3$

We make another assumption at this point. The SS is assumed to complete successful transmission of its CDMA code, in a maximum of 5 Backoff phases. Thus, the worst case of transmission will be four failures followed be a success. The final formula for the Backoff delay (BD) will be as follows:

$$BD \quad = p*\{AD_0 + t/2\}$$
$$+ ((1-p))*p*\{[AD_0 + t] + [AD_1 + t/2]\}$$
$$+ ((1-p)^2)*p*\{[AD_0 + AD_1 + 2t] + [AD_2 + t/2]\} \tag{7}$$
$$+ ((1-p)^3)*p*\{[AD_0 + AD_1 + AD_2 + 3t] + [AD_3 + t/2]\}$$
$$+ ((1-p)^4)*p*\{[AD_0 + AD_1 + AD_2 + AD_3 + 4t] + [AD_4 + t/2]\}$$

Here t is the time-out after which failure is assumed. So, we take half that value for success i.e. $t/2$.

3.3. Mathematical Derivation of the Overall IR Delay

By traversing the transition diagram (Figure 5) and multiplying the probabilities with the corresponding delays (Table 2), the total delay can be calculated. The resulting formula is as follows. The first part of the delay is in the loops 1-2-3-1 and 1-2-3-4-1. We call this D_{loop}. Then either success or abort occurs which is added to this part to get the final formula.

$$D_{loop} = 1*UL_M + 1*(BD + CDMA_{ST}) \tag{8}$$
$$+ a1 * (T3 + D_{loop})$$
$$+ a2 * (RSP + D_{loop})$$

Simplifying we get,

$$D_{loop} = \frac{UL_M + BD + CDMA_{ST} + a1*T3 + a2*RSP}{1 - (a1+a2)} \tag{9}$$

Now, the total delay involved can be represented using the formula given below.

$$D_{total} = D_{loop} \tag{10}$$
$$+ a3 * [RSP + CDMA_{IE} + RNG\text{-}REQ + (b1 + b2 + b3)*RSP]$$
$$+ a4 * RSP$$

(Here $b1 + b2 + b3 = 1$)

Substituting the expression for the delay in the loop into the formula for overall delay in IR, we get the following final formula.

$$D_{Total} = \frac{UL_M + BD + CDMA_{ST} + a1*T3 + a2*RSP}{1 - (a1+a2)}$$
$$+ a3*[RSP + CDMA_{IE} + RNG\text{-}REQ + RSP]$$
$$+ a4 * RSP \tag{11}$$

We define the Initial Ranging delay as the time taken by an SS to complete the IR scheme. Therefore, this is the time elapsed from the moment when an SS finds an IR opportunity to the moment when it receives an RNG-RSP message with a success status from the BS. This is done for the purposes of comparing the IR delay before and after the application of circularity. The IR delay process consists of the transmission delays of the request and response messages, the time needed for contention resolution, and the time needed by SS and BS to process the messages received.

$$\text{IR Delay} = \{\text{Time at which RNG-RSP with Success status is received}\} \qquad (12)$$
$$- \{\text{Time at which the first IR opportunity is found}\}$$

During the IR procedure, multiple SSs can send their request messages to the same BS simultaneously giving way to the occurrence of packet collisions. This leads to the contention resolution procedure being restarted with double the size of the backoff window. This leads to an increased IR delay. Thus, the contention resolution phase is the most affected by collisions among packets. We are interested in reducing the time spent by the SSs in competing with each other to send their request messages to the BS. This directly reduces the overall IR delay as well. With respect to the IR delay, we consider two approaches. Firstly, its reduction upon implementing circularity in IR is demonstrated. Then, the effect of applying circularity on BWR, IR and both BWR and IR upon the IR delay is studied.

4. ANALYSIS AND EVALUATION OF IR PERFORMANCE METRICS: SUCCESS RATIO

We define the IR success ratio as the ratio of the number of successfully completed IR procedures of various SSs to the sum of successfully completed IR procedures of various SSs and the number of retransmissions needed to be done as a result of the request packets timing out. This ratio is also directly affected by the collisions between RNG-REQ packets. This metric can be mathematically expressed as shown in the equations below:

$$\text{Success ratio} = \frac{\text{Successful attempts}}{\text{Total number of attempts}} \qquad (13)$$

$$\text{Success ratio} = \frac{\text{RNG-RSP with Success Status}}{\text{RNG-RSP with Success Status} + \text{RNG-REQ Expire}} \qquad (14)$$

Initially the success ratio is measured under 2 cases namely under the existing IR scheme and after implementing circularity for IR. Here, the effect of changing circularity values for delay and window control is also shown. Then, the success ratio is measured under 4 cases. They are existing IR scheme, with circularity for IR, with circularity for BWR and with circularity for both IR and BWR. In the next section we explain the paradigm of circularity that aims to reduce the number of collisions between request packets sent by various SSs.

5. PERFORMANCE ENHANCEMENT OF IR AND BWR

In IEEE 802.16 networks, the IR scheme is used by the SSs in order to acquire the timing offsets and the power adjustments from the BS, so that it can successfully transmit data packets. Although the mechanism is completely defined in the 802.16 standard, the performance of this mechanism is affected by the collisions between the RNG-REQ packets sent by different SSs.

In this section we propose an enhanced mechanism for IR, which incorporates the principle of circularity. Circularity is a novel technique whose objective is to reduce the number of collisions between the request packets in contention scenarios. It is defined as a number that allows us to identify specific groups of events or packets in the network [16].

The number of packets or events in one such group is equal to the circularity value. In each group, one of the packets or events is said to be circularity-satisfied. Here, we introduce certain control measures in case of circularity-satisfied packets and events. By doing this we achieve a decrease in the IR delay as well as an increase in the IR throughput. The circularity value is a positive integer. In order to identify the circularity-satisfied packets or events, we keep a count of the number of such packets or events. This count is common for all the SSs since we do not keep an individual counter for each SS. Whenever the value of this counter is a multiple of the circularity value, the packet or event is said to be circularity satisfied. If the counter is represented by k and the circularity value by c, then the mathematical representation for satisfying circularity is as follows

$$k \bmod c = 0 \tag{15}$$

The control measures taken are the following. Before sending the first RNG-REQ message or after sending its RNG-REQ packet if the SS does not receive a RNG-RSP message before a timeout, the RNG-REQ is said to have timed out. Then the SS doubles its backoff window and restarts the contention resolution procedure. The number of such expire events is counted. When an 'expire' event is circularity satisfied, the backoff window is doubled an extra time. By setting the backoff windows this way, in case of circularity satisfied expire events, the random numbers chosen by the different SSs will have lesser probabilities of being equal. This would mean that the backoff counters of the SSs would also have lesser probabilities of reaching zero at the same instant. Hence, the likelihood of collisions among the request packets decreases. After the requisite number of IR intervals is deferred, the SS is ready to send its RNG-REQ packet. We keep a count of such RNG-REQ packets as well. In the case of circularity satisfied RNG-REQ packets, we introduce a certain finite delay before the RNG-REQ packet is sent on the Initial Ranging Interval. Due to the delay introduced a particular request packet is sent a little later than it should have been. So, this sacrifice allows another SS to send its request packet in the meantime. This further reduces the probability of these packets colliding with each other.

In case of the BWR mechanism, the BWR packets are also kept count of. When the packet count of a BWR packet is a multiple of circularity, the packet is said to be circularity-satisfied. Such a BWR packet is discarded [17]. The events that occur in the IR intervals and the BWR contention slots of a particular frame affect the IR intervals and BWR contention slots of not only the next frame but also the frames after that. Consider the BWR contention slots of a particular frame 'U'. If there are many BWR packets sent successfully in a frame, then in the next frame 'U+1' there will be a larger number of uplink data bursts allocated for the data to be sent. This would of course be dependent on the availability of bandwidth in the next frame. A larger amount of data being sent implies fewer resources available on the uplink for contention based processes namely IR and BWR. Next consider the events in the IR intervals. After the successful transmission of RNG-REQ packets, the SS will complete the other stages of network entry and commence their BWR mechanisms. Thus, the IR process affects the activities of the BWR contention slots of the subsequent frames. Therefore, the activities of IR and BWR in a particular frame affect each other in the future. Hence, we try to establish a more efficient usage of the resources assigned by the base station for the contention-based processes. This is what circularity aims to do. Although our main focus remains the enhancement of the IR scheme, the implementation of circularity for BWR and its effect on IR is also dealt with in the next section.

6. ANALYSIS AND ENHANCEMENT OF BWR

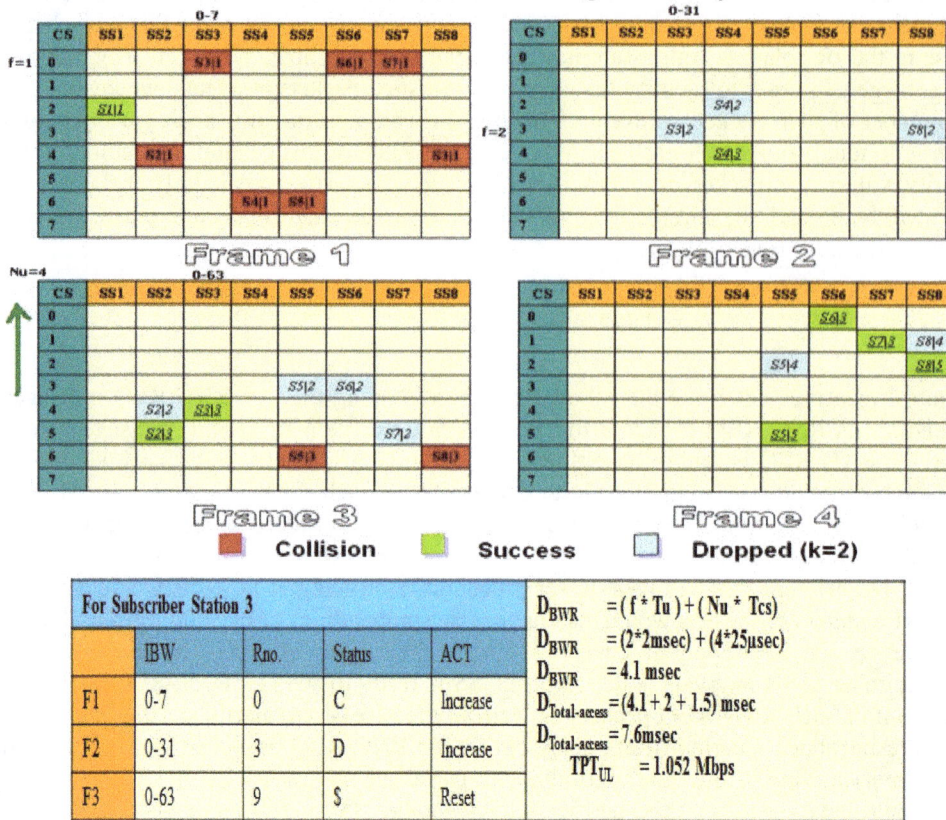

Figure 6. Frame Analysis of BWR with Circularity = 2

In order to explain the effects of the principle of circularity on the bandwidth request mechanism, we carry out a frame analysis that shows the events that occur during the mechanism. The frame analysis of BWR with the principle of circularity is as shown in the Figure 6.Let us explain the analysis taking SS1 as an example. First, SS1 tries to transmit its BWR in the first frame i.e. (F1) in the Contention Slot 2. Here, no other SS is transmitting its BWR and SS1 is able to transmit its request message without any collision. Next we consider SS2. SS2 tries to transmit its BWR in the first frame in CS4. But in this CS, SS8 is also sending its BWR. Hence these packets will collide which is denoted by C.

After the collision each SS, doubles its backoff window (if it has not reached a maximum already). Then they will select a number randomly from the new backoff window. The requisite number of BWR CS is deferred. The second BWR packets of SS2 and SS8 have been dropped, since they are circularity satisfied. This is denoted by D. The backoff window is doubled again. SS2 selects the number 10 and SS8 selects the number 11. With this value selected SS has a successful transmission of its BWR packet. But, SS8 again experiences a collision. The standard backoff procedure is continued in this case. Since the circularity value chosen is k = 2 for simplicity, every alternate request packet is dropped by each SS. Another form of frame analysis is shown in the following figure illustrating the events associated with each SS on a frame-by-frame basis. The performance metrics used to compare the BWR scheme with and without circularity are the access delay and throughput.

We define the access delay of the BWR scheme as follows [17].

$$D_{Total-access} = D_{BWR} + D_{gr} + T_{data} \tag{16}$$

Where D_{BWR} is the time spent in the contention process. This can be expressed mathematically as follows:

$$D_{BWR} = (f. T_U) + (N_U * T_{CS}) \tag{17}$$

After calculating the total access delay we compute the uplink throughput, for transmitting data of 1000 bytes, using the formula given below.

$$TPT_{UL} = \frac{Bytes\ sent\ *8}{D_{Total-access}} \tag{18}$$

7. SIMULATION STUDIES

The following sections describe in detail the simulation scenarios of various cases.

7.1. Simulation Setup

The simulations have been carried out using the Network Simulator 2 (ns-2) which is a discrete event simulator. We have added the WiMAX patch developed by the Advanced Network Technologies Division of the National Institute of Standards and Technologies [18]. The simulation script is written in the Tool Command Language (Tcl) [19], [20]. The model currently implemented is based on the IEEE standard 802.16-2004 and the mobility extension 80216e-2005 [21]. The model currently supports TDD. In this model, the uplink transmission follows only after the downlink occurs in each frame. The DL_MAP and UL_MAP messages determine the bursts of allocation and transmission opportunities for each station. The BS allocates slots that involve contention in the uplink direction. These slots are used in two cases:

- Initial Ranging request
- Bandwidth request

This model resolves contention by using a truncated binary exponential backoff procedure. The BS broadcasts the UCD messages that determine the window sizes. Also the BS decides the number of contention slots allocated in each frame.

The WiMAX control agent is used in the Tcl script in order to produce a detailed account of the activities in the network [21], [22]. The network configuration used is as follows. A single base station (BS) is considered. A sink node is considered that is attached through a wired link to the BS. The number of subscriber stations is simulated with BS. The performance metrics used are the delay, collision and throughput. The values for these metrics are calculated from the output file generated by the WiMAX control agent.

7.2. Implementation of Circularity

The basic network component objects in the data path are written in C++. The event scheduler is also scripted and compiled using C++. These compiled objects are made available to the OTcl interpreter through an OTcl linkage that creates a matching OTcl object for each of the C++ objects and makes the control functions and the configurable variables specified by the C++ object act as member functions and member variables of the corresponding OTcl object [23]. The backend files of the WiMAX module are coded in C++ are modified to implement the principle of circularity [24].

Table 3. List of important parameters that have been used during the simulation

channel Type	WirelessChannel
Radio Propagation Model	TwoRayGround
Network Interface Type	Phy/WirelessPhy/OFDM
MAC Type	802_16
Interface Queue Type	DropTail Priority Queue
Link Layer Type	LL
Antenna Model	OmniAntenna
Maximum Packets in Interface Queue	50
Routing Protocol	DSDV
BS coverage	20 meters
Simulation Time	50 seconds

In the case of Initial Ranging, we introduce two modifications namely delay control for the request packets and window control for the backoff Window. Firstly, a counter is kept for the RNG-REQ packets that are scheduled to be sent from the different SS. Whenever the counter value is an integer multiple of the circularity value, a finite delay is introduced. Secondly, a counter is kept of the number of RNG-REQ expire events that occur. Whenever this counter is a multiple of the circularity value, the Backoff Window is quadrupled (instead of the original doubling). This is done provided the Backoff Window does not exceed the designated maximum.

In the case of Bandwidth request, we selectively drop the request packets. A counter is kept for the BWR packets scheduled to be sent by the SSs. When this counter value is a multiple of the circularity value, the corresponding BWR packet is dropped.

7.3. Simulation Results

The following sections describe the simulation analysis and comparison of various cases for IR, BWR and both combined IR and BWR modules using ns-2 with different circularity values.

7.3.1. Simulation with IR

We describe the results of the simulations conducted using ns-2 [25], [26] in this section. In the first graph shown in Figure 7, we compare the delay incurred in the IR mechanism in the existing and enhanced scenarios. The circularity value used in selectively delaying the RNG-REQ packets is 3. The circularity value used in selectively doubling the backoff window an extra time is 5. The second graph compares the success-ratio of IR between the existing and improved IR schemes. There are three sets of Circularity values that are plotted in the graph shown in Figure 8 along with the existing IR scheme.

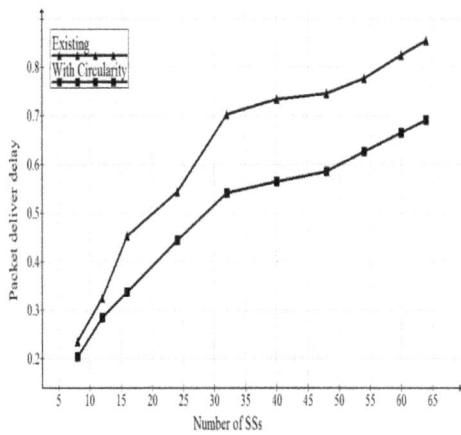

Figure 7.IR Delay comparison

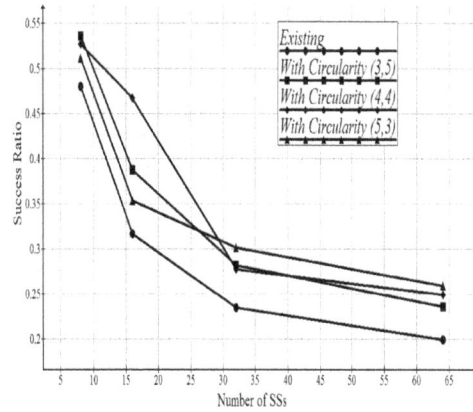

Figure 8. IR Success-Ratio comparison across three different sets of circularity valuepairs

Three circularity value pairs, used for delay control and window control, are considered. They are (3, 5), (4, 4) and (5, 3). The implementation of circularity in the IR scheme has decreased the delay incurred and increased the throughput. Within the improved scenario we observe the following. Around the 8 node scenario, for small network sizes, the circularity value pair (3, 5) gives higher throughput. Around 16 node scenario the circularity value pair (4, 4) gives higher throughput. For higher numbers of SSs (32 and 64) the circularity value pair (5, 3) gives higher throughput. From the above observation the following conclusions can also be made. With increasing number of SS, the circularity value controlling the delay must be increased. This implies that for increasing network sizes, the introduction of delay must be less frequent. With increasing number of SS, the circularity value controlling the window size must be decreased. This implies that for increasing network sizes, the doubling of window size must be more frequent.

7.3.2. Simulation with BWR

Now, we digress a bit and show the reduction in the time taken by SSs to complete their BWR mechanisms. This is achieved by applying the principle of circularity to selectively drop the BWR packets.

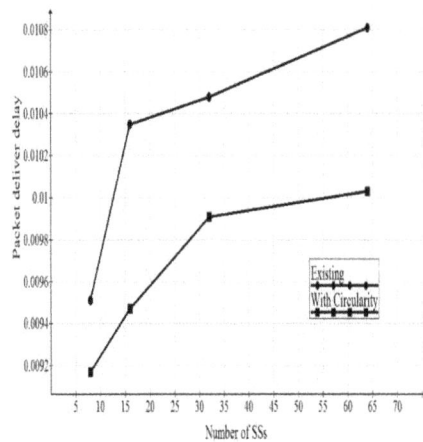

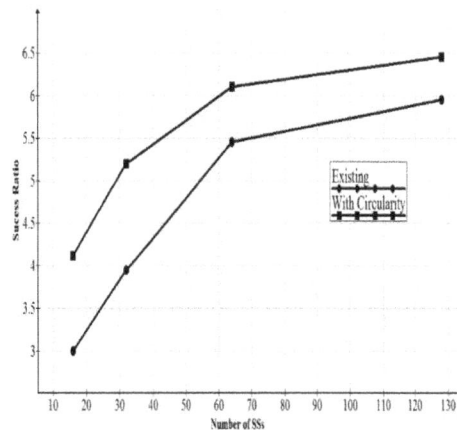

Figure 9.BWR Delay comparison with circularity using ns-2

Figure 10. BWR Delay comparison with circularity using generic code (C++)

This achieves a significantly more efficient utilization of the BWR contention slots that are available on the uplink. The graph showing this improvement follows. By altering Tcl code of BWR file,we can observe significant changes in access delay [27] with circularity for different test cases. Let us consider three different test cases in the following:

Case 1: Placing all Mobile Nodes (SS's) crowded.   Case 2: Placing Mobile Nodes (SS's) equidistant.

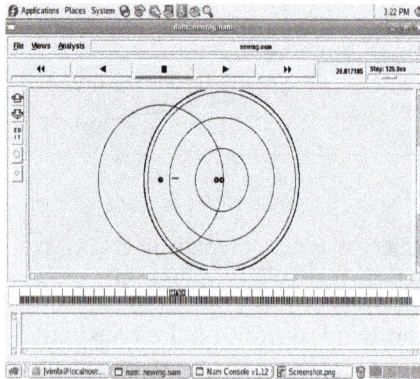   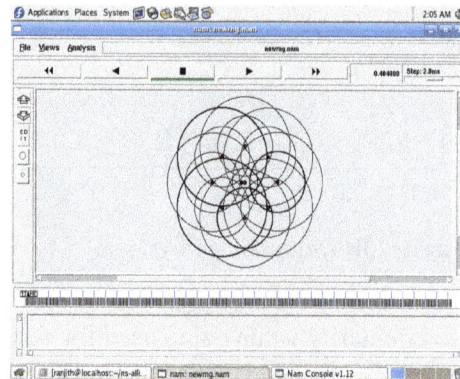

Figure 11.Placement of all mobile nodes   Figure 12.Placement of all mobile nodes
accessing from same point   accessing from equidistant point

Case 3: Mobile Nodes (SS's) on the circumference.

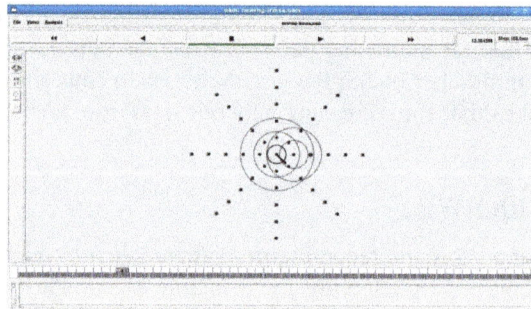

Fig.13.Circumferential Topology with no motion

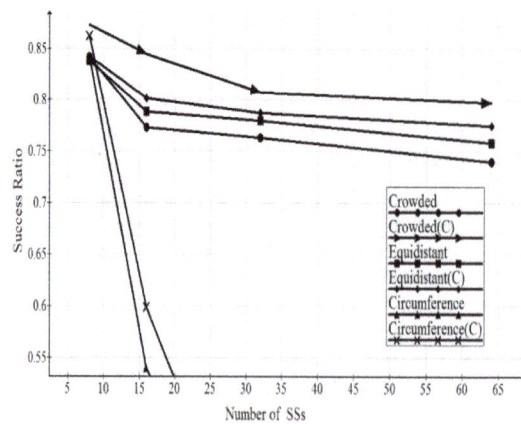

Figure 14.Comparison of various topological scenarios.

Simulation results of delay and throughput for the Case 1 wherein the SSs are densely placed in the same location are depicted in the Figures 9 and 10. The Figures 11, 12 and 13 illustrate different topological scenarios consider for further simulation. The comparative results are presented in Figure 14. The enhanced performance is evinced in all cases while adopting the principle of circularity for improving the contention resolution process. In circumferential topology, the coverage area between BS and SSs expands farther away as network size increases. This results in drastic drop in the throughput.

7.3.3. Simulation with IR and BWR

Due to the influence of contention in the IR and BWR mechanisms on each other, a more efficient collision avoidance mechanism will result in a lesser delay and a higher success-ratio for the IR mechanism. This is proved by the following graphs. Four cases are considered here. Firstly in Figure 15, the delays for the existing IR scheme and the IR scheme with circularity are plotted. Then the delay with circularity implemented for BWR and both IR and BWR are shown.

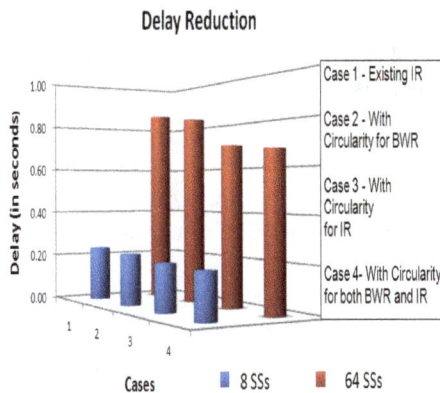

Figure 15. Variation of IR delay across four cases of circularity implementation

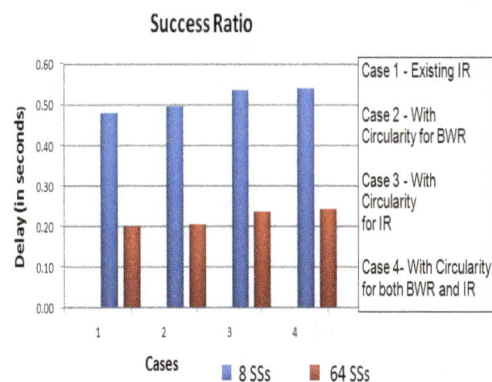

Figure 16. Variation of IR success ratio across four cases of circularity implementation

In Figure 16, the same four cases are considered for the success-ratio of the IR mechanism. From the third and fourth set of results, the influence of the efficient use of IR intervals and the BWR contention slots on the IR delay and the success ratio of RNG-REQ packets are established. Irrespective of whether circularity is applied to IR or BWR, the IR performance improves. Also better IR performance is observed when circularity is applied to IR mechanism. Finally, the performance of IR is best when we implement circularity for both IR and BWR.

8. CONCLUSION

The Initial Ranging scheme is a critical part of the network entry procedure. The delay incurred during the mechanism is mainly a function of the delay incurred in the backoff procedure that is at the center of IR. Due to collisions among RNG-REQ packets, the backoff procedure may be repeated many times leading to higher IR delay and thereby increased time to complete the network entry procedure. The concept of Markov process is used in analyzing the IR mechanism. From the above observation we can also conclude the following: 1) with increasing number of SS, the circularity value controlling the delay must be increased. This implies that for increasing network sizes, the introduction of delay control must be less frequent. Similarly, 2) with increasing number of SS, the circularity value controlling the window size must be decreased. This implies that for increasing network sizes, the doubling of window size must be more frequent.

Bandwidth Request (BWR) is the process wherein the communication takes place between SSs and BS for uplink bandwidth allocation. We introduce the concept of circularity aiming to reduce collisions among the BWR packets and make better use of the contention slots. We carry out a frame analysis that shows the events that occur during the mechanism. We have proved that the selective dropping of BWR packets decreases the number of collisions owing to improved mechanism of collision avoidance. This directly leads to decrease in the total access delay required by an SS to complete its BWR procedure. The throughput of the data sent is thereby considerably increased.

Furthermore, we introduce the circularity concept in both IR and BWR module combined at the same time. This will reduce the time required by the SSs to successfully transmit their request packets. Hence both IR and network entry get completed in lesser time and also thecontention process in BWR mechanism is resolved faster. Thus, reduction in number of collisions increases the success ratio of request packets sent. We have also analyzed and compared the effects of the circularity value with IR, BWR and combined IR and BWR modules on the success ratio.

ACKNOWLEDGEMENTS

The authors sincerely thank the authorities of Supercomputer Education and Research Center, Indian Institute of Science for the encouragement and support.

REFERENCES

[1]   IEEE Standard for Local and Metropolitan area networks, (2004) Part 16: *Air Interface for Fixed Broadband Wireless Access Systems*; IEEE Computer Society and IEEE Microwave Theory and Techniques Society.

[2]   IEEE Standard for Local and metropolitan area networks, (2005) Part 16: *Air Interface for Fixed and Mobile Broadband Wireless Access Systems, Amendment 2: Physical and Medium Access Control Layers for Combined Fixed and Mobile Operation in Licensed Bands*; IEEE Computer Society and IEEE Microwave Theory and Techniques Society.

[3]   Chakchai So-In, Raj Jain, and Abdel-Karim Al Tamimi, (2010) "Capacity Evaluation for IEEE 802.16e Mobile WiMAX," Journal of Computer Systems, Networks, and Communications (JCSNC),   Special issue on WiMAX, LTE, and WiFi Interworking, Vol. 1, No. 1.

[4]   Chakchai So-In, Raj Jain, Abdel Karim Al Tamimi, (2010) "A Scheduler for Unsolicited Grant Service (UGS) in IEEE 802.16e Mobile WiMAX Networks," IEEE Systems Journal, V-4, No. 4.

[5]   J. Y. Kim and D. H. Cho, (2007) "Piggybacking Scheme of MAP IE for Minimizing MAC Overhead in the IEEE 802.16e OFDMA Systems", In IEEE 66th Vehicular Technology Conference (VTC-2007 Fall), pp284-288.

[6]   Bhandari, B. N., Ratnam, K., Raja Kumar, and Maskad, X. L, (2005) "Uplink Performance of the IEEE 802.16 Medium Access Control (MAC) Layer Protocol", ICPWC, Hong Kong.

[7]   Ahmed Younus, (2009) "WiMAX – Broadband Wireless Access", Tech. Univ. of Munich, Germany.

[8]   J. He, K. Guild, K. Yang, and H.-H. Chen, (2007) "Modelling Contention Based Bandwidth Request Scheme for IEEE 802.16 Networks", IEEE Communications Letters, 11(8):698-700.

[9]   S. M. Oh and J. H. Kim, (2005) "The Analysis of the Optimal Contention Period for Broadband Wireless Access Network", in Third IEEE International Conference on Pervasive Computing and Communications Workshops, pp215-219.

[10]   J.G. Andrews, A.Ghosh. and R. Muhamed, (2007) "Fundamentals of WiMAX Understanding Broadband Wireless Networking", Pearson Education Inc.,One Lake Street Upper Saddle River, NJ07458, USA.

[11]   E. Nuaymi (2007) "WiMAX Technology for Broadband wireless Access" John Wiley & Sons Ltd, the atrium. Southern gate, Chichester, West Sussex PO198SQ, England.

[12]   J. Yan and G. S. Kuo, (2006) "Cross-layer Design of Optimal Contention Period for IEEE 802.16 BWA Systems", In IEEE International Conference on Communications, Vol. 4, pp1807-1812.

[13]   Sheldon M. Ross, (2000) "Introduction to Probability Models", 7th Edition. Academic Press, New York.

[14]   WiMAX Forum, Krishna Ramadas and Raj Jain, (2008) "WiMAX System Evaluation Methodology"   No. 2.1.

[15]   Lidong Lin, WeijiaJia, Bo Han and Lizhuo Zhang, (2007) "Performance Improvement using Dynamic   Contention Window Adjustment for Initial Ranging in IEEE 802.16 P2MP Networks", Proceedings of the IEEE Wireless Communication and Networking Conference, Hong Kong.

[16]   Mohammad Z. Ahmad, DamlaTurgut, and R. Bhakthavathsalam, (2006) "Circularity based Medium Access Control in Mobile Ad hoc Networks", 5th International Conference on AD-HOC Networks & Wireless (ADHOCNOW'06), Ottawa, Canada.

[17]   R. Bhakthavathsalam and Khurram J. Mohammed, (2009) "Analysis and Enhancement of the BWR mechanism in MAC 802.16 for WiMAX"; Proceedings of the 6th Annual IEEE Consumer Communications & Networking Conference, Las Vegas, Nevada, USA.

[18]   National Institute of Standards and Technologies, (2011)"*Seamless and Secure Mobility Tool Suite*", Advanced Network Technologies, http://www.antd.nist.gov/seamlessandsecure/doc.html.

[19]   M. Greis, (2010) Tutorial for the Network Simulator NS2. [Online]. Available: http://www.isi.edu/nsnam/ns/tutorial/

[20]   D. Libes, (1993) "A debugger for Tcl applications," in Tcl/Tk Workshop, [Online]. Available: http://expect.nist.gov/tcl-debug/tcl-debug.ps.Z

[21]   The Network Simulator – ns-2. (2012) [Online]. Available: http://www.isi.edu/nsnam/ns/

[22]   The Network Simulator Wiki (2010) [Online]. Available: http://nsnam.isi.edu/nsnam/ index.php/

[23]   Berkeley Continuous Media Toolkit. *OTcl Tutorial*. (2012) [Online]. Available: http://bmrc.berkeley.edu/research/cmt/cmtdoc/ otcl/

[24]   An AWK primer.(2011) [Online]. Available: http://www.vectorsite.net/tsawk.html

[25]   Bazzi, A. Leonardi, G. Pasolini, G. Andrisano, O. DEIS, (2009) "IEEE802.16e Simulation Issues" Napa Valley, CA.

[26]   Sanfilippo, (2012)   "*An introduction to the Tcl programming language*" [Online]Available: http://www.invece.org/tclwise

[27]   The   GDB   Developers,(2012)"*GDB: The   GNU   project   debugger*"   [Online] Available:http://www.gnu.org/software/gdb

8

Hybrid Scheme for Discovering and Selecting Internet Gateway in Mobile Ad Hoc Network

Shahid Md. Asif Iqbal[1] and Md. Humayun Kabir[2]

[1]Department of Computer Science & Engineering, Premier University, Chittagong, Bangladesh
asifcsep@yahoo.com
[2]Department of Computer Science and Engineering, Bangladesh University of Engineering and Technology, Dhaka, Bangladesh
mhkabir@gmail.com

ABSTRACT

Global connectivity to Mobile Ad Hoc Network (MANET) is necessary to access the Internet services from the MANET. Nodes in a MANET that connect it to the Internet are called Internet gateways. Internet gateways need to be discovered and selected in an appropriate way to deliver more packets to the Internet and reduce end-to-end delay. Currently, there are proactive, reactive, and hybrid schemes to discover and select Internet gateways in MANET. However, these schemes do not scale well with the number of nodes, traffic load, and speed of the nodes in MANET. To make it scalable, we propose a new gateway discovery and selection scheme. In our scheme, the gateways advertise gateway advertisement messages only on-demand. Moreover, it contains the advertisements within a limit in order to make our scheme scalable. We also consider the interface queue length and the total number of neighbors along a route in addition to the hop count to bypass the loaded and dense route to the gateways in order to reduce the delay and packet loss. Simulation results show that our scheme scales well with the number of nodes, traffic load and the speed of the nodes in MAENT compared to that of other schemes. It also confirms that our scheme has less delay and packets drop than that of other schemes.

KEYWORDS

Internet, MANET, Integration, Internet Gateway Discovery, and Gateway Selection.

1. INTRODUCTION

A Mobile Ad Hoc Network (MANET) is formed by a group of mobile nodes without the aid of any centralized administration or established infrastructure. A pair of mobile nodes may communicate with each other either directly or indirectly with the help of the intermediate nodes. Since these kinds of networks are very spontaneous and self-organizing, many useful applications such as multimedia streaming, collaborative work, information dissemination and jungle telemetry can be supported by these networks and that's why they are very demanding in commercial arena specially in the emergency services like hospitals, ambulance, police and military applications etc.

In future, the Internet is likely to be different from its present state because mobile devices with various computational resources will dominate it. Wireless communication technology and the Internet are developing so quickly that there are numerous mobile devices around us and multiple wireless networks are serving these mobile devices all the time. A MANET is generally considered as a stand-alone network i.e. communication is only supported among the nodes within the ad hoc domain. This stand-alone nature limits the applicability of MANET to the scenarios those require external connectivity. Integration of MANET and Internet can provide global connectivity to MANET so that it no longer remains stand-alone. This integration allows mobile users in MANET to access the popular Internet applications such as e-mail, chat, instant messaging, file transfer etc. The integration expands both MANET and the Internet coverage range.

Integration of MANET with the Internet has recently become an active research area. To access the Internet from a MANET a subset of its nodes must have the interfaces to connect to the Internet directly. These nodes work as the Internet gateway, which facilitates other nodes to communicate outside the MANET. There might be multiple Internet gateways in a MANET. A mobile node in a MANET may be multi-hop away from the Internet gateways. In this case, the node has to use the Internet gateway through other intermediate nodes.

When a mobile node in a MANET wants to access the Internet, it needs to discover the available Internet gateways and selects the best one among them if multiple gateways are found. Therefore, it needs an efficient Internet gateway discovery and selection scheme that achieves high throughput, low delay and less network-overhead. Two types of schemes, reactive and proactive, have been proposed to discover and select Internet gateways in MANET. In proactive schemes [1-9], Internet gateways periodically broadcast gateway advertisement messages in the MANET. Each node that receives the advertisement message forwards the advertisement to other nodes until the message is flooded over the whole network. These schemes cost heavy routing load since the gateway advertisements are broadcasted periodically throughout the entire ad hoc network even if there is no such demand from the nodes in the MANET. However, the proactive schemes are blessed with higher rate of successful delivery and lower delay. In reactive schemes [1-3] [9-12], a mobile node broadcasts a gateway discovery message to discover Internet gateways in the network. Whenever a gateway receives the discovery message, it unicasts a gateway advertisement message back to the requestor. These schemes suffer from higher delay and lower packet delivery ratio since the nodes have to send a gateway discovery message every time they need a gateway. Reactive schemes scale poorly regarding the number of sources willing to access the Internet. Few research works [9] [13-19] proposed hybrid gateway discovery schemes where the dissemination of gateway advertisements is kept limited to a small area by setting appropriate Time to Live (TTL). Nodes outside the TTL coverage area reactively find their gateways. The performance of these schemes degrades if TTL is not adapted properly. Most of the existing hybrid schemes [9] [13-14] [16] do not adjust TTL value dynamically.

Gateway selection scheme selects the best gateway when it receives multiple gateway advertisements from multiple gateways. Gateway selection schemes proposed in [1-3] [5] [7] [9] [13-18] use hop count only to select a gateway. In these schemes, all the nodes always select the nearest gateway, a gateway may become a bottleneck under heavy traffic load and there is no remedy for this problem.

To deal with the problems in existing Internet gateway discovery and selection schemes we propose a new hybrid gateway discovery scheme where gateways will act reactively, however, broadcast a gateway advertisement message when they receive a gateway discovery message from a mobile node. The TTL of the gateway advertisement message will also be set to a value equal to the distance of the gateway from the requestor. Each mobile node will configure its gateway after receiving the gateway advertisement message. In our scheme, a node selects a gateway that promises optimal performance, after receiving the advertisement messages from multiple gateways. While selecting the best gateway, the node will consider the interface queue size and the total number of neighbors of each node along the route in addition to the hop count. We consider the number of packets waiting in the interface queue of a mobile node as its interface queue size. The use of interface queue size in the selection of a gateway, allows us to redirect a mobile node from a heavily loaded gateway to a less loaded one and the inclusion of the total number of neighbors of each node helps us to avoid a crowded area to reach the gateway.

The rest of the paper has been organized as follows. In Section 2 we review the current solutions for Internet gateway discovery and selection in MANET. Section 3 depicts our new hybrid Internet gateway discovery scheme. We also introduce the new metric used in the gateway selection scheme in this section. Simulation setup and analysis of simulation results comes in Section 4 and finally in Section 5 we conclude our paper with some future research guidelines.

2. RELATED WORKS

During the last decade, many works have been devoted to the study of ad hoc routing protocols, but the decade lacks adequate works to provide Internet connectivity to the nodes in MANET. Since Internet has made information more available and easier to access, the desire for having a MANET connected to the Internet is increasing. Typically, several gateways in a MANET connect the network to the Internet. The rest of the nodes discover the available gateways and select the best one among them.

2.1. Internet Gateway Discovery Schemes

Recently the issue of Internet connectivity to MANET has been addressed by [1-24]. MIPMANET [3] was designed to provide nodes in the ad hoc networks with access to the Internet and the mobility services of IP. A foreign agent (FA) in MIPMANET [3] acts as an access point and provides Internet connectivity to an entire ad hoc network. It uses a single IP address as a care-of-address and a reverse tunneling to provide Internet access to the nodes. Each FA in the MANET broadcasts foreign agent advertisement messages periodically. Mobile nodes in the network use ad hoc on-demand distance vector (AODV [25]) routing protocol for routing within the MANET. FAs have the MIPMANET Internetworking Unit (MIWU) that is inserted between the FA and the ad hoc network. MIPMANET uses MIPMANET Cell Switching (MMCS) algorithm to handover between foreign agents. Belding-Royer et al. [22] proposed Mobile IP for IPv4 ad hoc networks using AODV routing protocol. In that proposal, a node first has to determine the location of the destination node before it starts sending packets to that destination. Here, a FA unicasts a route reply (F-RREP) message when it receives a FA discovery message from a mobile node. Mobile nodes use the F-RREP messages to determine the location of the destination nodes. It is capable of routing packets to FA using default route. A disadvantage of this proposal is that, a mobile node has to know that the destination of a packet is not within the ad hoc network before sending it to the FA, which in turn increases the delay for connection setup.

In [1], the authors discussed the technique to provide global Internet connectivity to IPv6 MANET environment using on-demand routing. The paper proposed two Internet gateway discovery schemes: proactive gateway discovery scheme using periodic gateway advertisement messages from the gateway and reactive gateway discovery scheme by flooding gateway discovery messages from the nodes. Lee at el. [13] proposed two gateway advertisement schemes based on the observation of traffic and mobility pattern of nodes to avoid unnecessary routing overhead in MANET. However, the scheme relies on source routing protocol that limits the applicability and scalability of the solution.

In addition to the reactive or proactive gateway discovery schemes [1-12] there are some research works [9] [13-19] that proposed hybrid gateway discovery schemes. In the hybrid schemes, the time-to-live (TTL) value of the gateway advertisements is kept limited to certain boundary in order to contain the proactive discovery within an optimum range. These schemes are mainly designed to minimize the disadvantages of proactive and reactive schemes i.e. to provide good connectivity and low overhead. However, these schemes require some intelligent adaptation of the TTL value. In [19] authors proposed a load-adaptive access gateway discovery protocol that defined a proactive range for the gateway advertisement which is dynamically adjusted according to the changing network conditions. Nevertheless, the gateway advertisement scheme is effective when there are only fixed sized packets in the network. Here the authors used the network size and the number of nodes in the network to compute the initial proactive range, which is unlikely because there may be no good technique to know the size and the number of nodes in a MANET.

2.2. Internet Gateway Selection Schemes

If a node discovers multiple gateways then it has to decide which one is to use. Majority of current gateway selection schemes [1-3] [5] [7] [9] [13-18] [22] use hop count to select the best gateway, and they always select the nearest gateway with the hop count metric. If all the mobile nodes always select their nearest gateway then the nearest gateway may become bottleneck under heavy traffic load, also there might be congested nodes along the route to the gateway. That is, hop count based selection schemes choose a gateway that might have less capacity and difficult to reach. As a result, network performance degrades with the hop count metric.

Few research works [4] [6] [8] [10-12] [19-20] considered traffic load in addition to the hop count to select the best gateway. Each of these research works treated traffic load differently than the others. Kumar et al. [4] considered the number of packets waiting in the interface queue of the nodes to select a gateway. Khan et al. [6] considered the number of packets waiting in the routing queue of the nodes to select a gateway. However, both of these works converted the number of packets into equivalent hop count without proper justification, which may not provide the actual traffic load information. Le-Trung et al. [10] proposed a hybrid metric for Internet gateway selection that provides load-balancing of intra/inter-MANET traffic. However, the selection scheme introduces extra routing load and requires high processing power consumption to compute the hybrid metric. Li et al. [11] considered the speed of the nodes along with node's available energy and traffic load to select a gateway. Zhanyang et al. [12] also considered the speed of the nodes to compute the gateway selection metric. Nevertheless, obtaining the speed of a node impose additional cost which may limit the applicability of the work. QoS-enabled access gateway selection scheme proposed in [19] considered the packet arrival rate of a gateway in an interval as the traffic load. It uses a Decision Function (DF) that considered the traffic load and hop count to select a gateway. In this case, each intermediate node needs to piggyback its load information periodically on data packets, which increase the header size of the data packets. In [20], the authors proposed a gateway selection scheme based on hop count, gateway load and path quality, and make use of a hybrid search approach which is based on orthogonal genetic algorithm and sensitivity analysis. The authors have used the maximum packet queue size, average packet queue size and an index α to compute the gateway load. However, the computation of average packet queue size depends on the periodical gateway advertisement and better average can only be obtained for smaller advertisement interval. The authors did not talk about how to select the value of α either. In [20], the authors used the variance in arrival times of periodical gateway advertisement broadcast messages in order to evaluate the quality of the path between mobile nodes and the gateway. However, the computation of the variance needs an intelligent selection of a history window in order to express how long history needs to be considered when calculating the mean value and variance. This makes their selection scheme effective for periodical gateway advertisement only with small advertisement interval. Nevertheless, periodical gateway advertisement with small advertisement interval results in tremendous routing load in the network.

3. PROPOSED INTERNET GATEWAY DISCOVERY AND SELECTION SCHEME

In this section, we describe our proposed Internet gateway discovery and selection scheme for MANET. At first, we present the network architecture that our scheme is based on. After that, we describe our Internet gateway discovery scheme. We also show the computation of the metrics that are used in our Internet gateway selection scheme.

3.1. Network Architecture

We assume a regular MANET consists of two types of nodes. One type of nodes has Internet connectivity, we call them Internet gateways, and the other type of nodes that don't have Internet connectivity but they can access the Internet through the Internet gateways. We call this second type of nodes simply, mobile nodes.

We assume all the nodes in our MANET have equal transmission range. Nodes can communicate directly with each other if they fall in each other's transmission range. Nodes who are not within each other's transmission range can also communicate indirectly via one or more intermediate nodes. Nodes can join or leave the network anytime. Nodes are free to move in any direction. We did not impose any Internet bandwidth limitation on the Internet gateways

Internet gateways in our MANET can access the Internet themselves. However, the mobile nodes have to access the Internet through an Internet gateway. For this reason, mobile nodes have to discover the gateways first. We describe our gateway discovery scheme in Section 3.2. If multiple gateways are discovered by a mobile node, the best gateway must be selected to access the Internet. We describe our gateway selection scheme in Section 3.3. Any MANET routing protocol such as AODV [25], OLSR [26] and DSR [27] can be used to route the packets within our network.

3.2. Internet Gateway Discovery

When a mobile node in the MANET wants to access the Internet, at first it has to find a gateway. Like [4] [6] [9], a mobile node in our gateway discovery scheme looks in its routing table to find a default route i.e. a route to a gateway. If the mobile node finds a default route, it uses the route to send packets to the gateway i.e. to the Internet.

However, if the mobile node does not find a route to a gateway in its routing table, we propose it to start a gateway discovery process by broadcasting a gateway discovery (GWDSC) message in the MANET. While broadcasting the GWDSC message, we propose the requesting mobile node to set an initial time to live (TTL) value for the message and start a timer to wait for the reception of the gateway advertisement message from the gateways. Figure 1 shows the format of the GWDSC message.

0	8			12	24 31
Type	J	R	G	I	Reserved	Hop Count
RREQ_ID					
Destination IP Address					
Destination Sequence Number					
Originator IP Address					
Originator Sequence Number					

Figure 1. Format of GWDSC messages in our scheme

In our scheme, upon receipt of a GWDSC message, an intermediate node creates a reverse route entry for the requestor in its routing table and forwards the GWDSC message to its neighbors. In this way, a GWDSC message reaches one or more Internet gateways in the network if there is any.

We propose an Internet gateway to broadcast a gateway advertisement (GWADV) message when triggered by a GWDSC message. We also propose to set the TTL value of the GWADV message equal to the distance of the gateway from the requesting mobile node. In our scheme, we control the TTL value of the GWADV message to contain the dissemination of the GWADV message to a certain range, which helps to reduce the routing overhead to an extent. We allow gateways to broadcast GWADV messages only in response to GWDSC messages in order to avoid unnecessary flooding of GWADV messages in the network.

In addition to the conventional fields, we have added two new fields in the GWADV message header. We name these new fields Q and N respectively. We use the Q field to represent the total interface queue size of nodes along a route from a gateway to a mobile node. We use N field to represent the total number of neighbors of the nodes along a route from a gateway to a mobile node. We use the Hello messages of ad hoc routing protocols to obtain the neighbor information of a gateway or a mobile node.

We propose an Internet gateway to populate these two fields before flooding a GWADV message. We also propose intermediate nodes to update these two fields while forwarding the message to the next nodes. The modified structure of a gateway advertisement message header in our scheme is given in Figure 2.

0	8	19	24 31
Type	Reserved	Pref. Sz.	Hop Count
Broadcast_ID			
Destination IP Address			
Destination Sequence Number			
Source IP Address			
Lifetime			
Q			
N			

Figure 2. Format of GWADV Message in our scheme

Upon receipt of a GWADV message, we propose a mobile node to decrement the TTL first and to configure the corresponding gateway if it does not have a gateway configured yet. In this way, more nodes in the network will have the opportunity to configure their gateway without broadcasting a GWDSC message, i.e., our scheme will reduce the GWDSC message broadcast to a significant level. Mobile nodes that already have their gateway configured should reconfigure the gateway if the corresponding gateway seems better. A GWADV message is forwarded to the neighbors if the TTL value is not zero. In this way, we allow the GWADV message to reach to the requesting mobile node. Therefore, in our scheme, a GWADV message helps not only the requesting mobile node but also the other nodes in the network to configure their gateway. As a result, our proposed scheme helps a mobile node in a MANET to hand off to a better gateway even before its current Internet connection is broken.

However, if the requesting mobile node does not receive any GWADV message before the timer expires, we propose the node to broadcast a new GWDSC message with an increased TTL value. We propose the requesting mobile node to increase the TTL value linearly. We increment the TTL value linearly to experience less routing overhead (GWDSC messages). We allow this process to continue until either the requesting mobile node receives a GWADV message or it broadcasts a GWDSC message with a pre-defined maximum TTL value.

Thus, our gateway discovery scheme consists of on-demand GWDSC messages like reactive scheme, broadcast of GWADV messages like proactive scheme and limited TTL value for GWADV messages like hybrid scheme. That is, our scheme combines the bests of the three conventional Internet gateway discovery schemes and can provide efficient and faster discovery of Internet gateways.

3.3. Internet Gateway Selection

We propose a new composite metric to select the best gateway when a mobile node receives multiple gateway advertisement messages from multiple gateways; we call this new metric **gateway-cost (gc)**. Our metric **gc** is composed of three factors: hop count, interface queue size and total number of neighbors.

Like [1-3] [5] [7] [9] [13-18] [22], we consider hop count to select the best gateway. It denotes the number of nodes or routers between a mobile node and an Internet gateway. This factor allows a mobile node to reach the Internet using minimum number of hops which facilitates the rapid convergence and resource thriftiness of the network.

We consider the interface queue size of each node along a route to a gateway. Interface queue size of a node denotes the number of packets waiting in the interface queue of that node. If the size of the interface queue of each node along a route to a gateway is less, then more packets can be sent to the Internet using that route and the packets will have to wait less. Thus, we consider interface queue size of each node to allow fair distribution of the network load among the gateways and congestion prevention in the network.

We consider the total number of neighbors of each node along a route to a gateway. This factor helps a mobile node to select a gateway whose path is least dense. A least dense path is more likely to have least contention and best to use to reach the gateway. As far as we know, nobody used this factor to select a gateway in a MANET before us.

Whenever a node **p** in a MANET receives a GWADV message from a gateway **q**, we propose it to calculate **gc** using eq. (1):

$$gc_q = hc_q + \frac{Q}{Q+1} + \frac{N}{N+1} \qquad (1)$$

$$q \in V_{GW} \qquad Q = \sum_{i=1}^{hc_q} int\_q\_size_i \qquad N = \sum_{i=1}^{hc_q} n_i$$

Where V_{GW} is the set of Internet gateways present in the network, hc_q is the number of hops from **q** to **p**, $int\_q\_size_i$ represents the interface queue size of node i along the route from gateway **q** to node **p**, n_i represents the number of neighbors of node i along the route from gateway **q** to node **p**.

When a mobile node receives multiple GWADV messages from multiple gateways, we select the gateway with the lowest **gc**.

We give more emphasis on the hop count because it is always better to select a shorter route to minimize network delay and to optimize network resource usage. A packet routing through a shorter path also have better chance to face less network adversaries, such as bit error and congestion. Although the queue size and the number of neighbors along the route help us to avoid the gateways having bad route to reach, these two are actually less significant factor compared to hop count. Thus, if the two factors are kept intact like the hop count in the computation of the metric **gc**, then our selection scheme may choose a gateway which is not closest in terms of hop count. As a result, a mobile node in a MANET has to travel a longer route to reach an Internet gateway in the MANET. A longer route not only increases delay or consumes network bandwidth and node energy but it also involves more intermediate nodes to forward packets to an Internet gateway. A route to a gateway with higher number of intermediate nodes has better chance to suffer from more congestion and collision compared to that of smaller routes. Consequently, this fact may cause more packets drop and route re-discoveries in the network. For this reason, we give less emphasis on these two factors. To do so, we individually adjust these two factors so that they can contribute positively in the computation of the gateway selection metric **gc** but their individual contribution always remains less than 1. Therefore, our metric **gc** selects the gateway whose path is not only less loaded and less dense but also shortest.

4. PERFORMANCE EVALUATION

To evaluate the performance of our proposed Internet gateway discovery and selection scheme, we implemented our scheme in ns-2 [28] network simulator and compared the results with that of the proactive, reactive and hybrid schemes that were proposed in [29]. We also modified the MANET routing protocol AODV [25] to route packets between a gateway and a mobile node.

4.1. Performance Metrics

We compare all the Internet gateway discovery and selection schemes based on three performance metrics namely Internet Packet Delivery Ratio, Average End-to-End Delay, and Normalized Control Overhead. These are the standard performance metrics that are also used by many research works [4] [6-12] [19] to evaluate Internet Gateway Discovery and Selection Schemes.

The Internet Packet Delivery Ratio (IPDR): IPDR is defined as the ratio between the total number of data packets received by the corresponding destination hosts in the Internet and the total number of data packets sent to the Internet by all the mobile nodes in the MANET.

The Average End-to-End Delay: It is defined as the average time needed to send a data packet from a node to a host in the Internet. It is computed in milliseconds (ms).

The Normalized Control Overhead (NCO): NCO is defined as the ratio between the total number of AODV messages transmitted by the nodes in MANET and the total number of data packets received by the hosts in the Internet.

We vary the number of nodes in MANET from 10 to 30 to see the network behavior under different traffic load. The number of neighbors of each node also varies with the number of nodes in the MANET. We vary the speed of the nodes from 2 to 30 m/s which allows us to compare the performance of the schemes in different speeds, such as walking speed (2 m/s), downtown driving speed (10 m/s), suburban driving speed (20 m/s), and highway driving speed(30 m/s) [13].

4.2. Simulation Setup

This section describes the network scenario, the movement model, the communication model, and the simulation parameters that we have used in our study.

4.2.1. Scenario

Like [11] [14] [19-20], our simulated network is spanning in a standard area of 1000x1000m^2. Each mobile node in our simulation has a wireless transmission range of 250 meter, which is the standard range and also used by the other research works [4] [6] [9] [10] [11] [15] [19-21]. This transmission range ensures no network partitioning.

We have considered 4 Internet gateways in the MANET in our simulation scenarios in order to load balance the Internet traffic. We assume a higher Internet bandwidth for gateways compared to that of the MANET nodes. We set the Internet bandwidth of each gateway to 10 Mbps.

We ran our simulations for 500 units of simulation time. According to our observation, 500 units of simulation time is high enough to see the steady behavior of the network in different scenarios. The seed time for each node to send data packets is considered 0.5 units of the simulation time. This seed time confirms that all the schemes start their gateway discovery process before the nodes start sending the data packets to the Internet. A screenshot of a simulation scenario is given in Figure 3. In the figure, the red-colored hexagonal nodes represent the gateways, the blue-colored square nodes represent the Internet hosts and the green-colored circular nodes represent the mobile nodes.

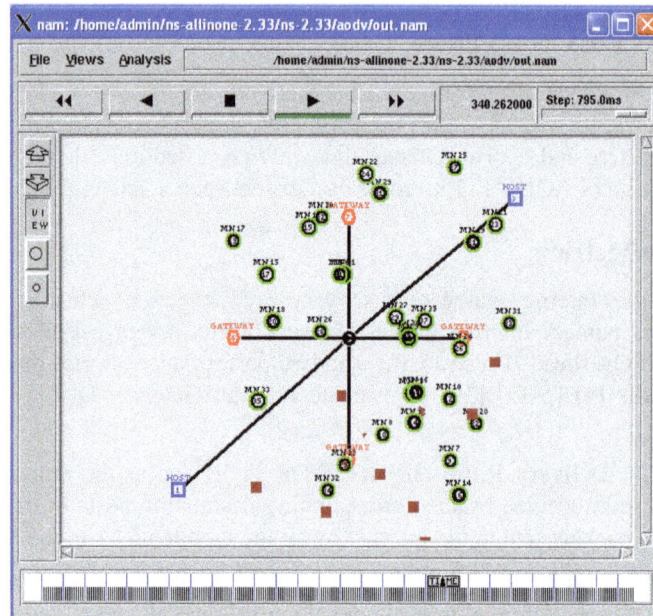

Figure 3. Screenshot of a Simulation Scenario

4.2.2. Movement Model

We used the Random Waypoint Movement Model [30] as the mobility model for our simulation. It is the benchmark mobility model that has been used in many research works [3-4] [6-9] [11-21] in order to evaluate network protocols in MANET. According to this model, a mobile node remains stationary for a certain period called *pause time*. After the *pause time* is over the node selects a destination randomly and moves to that destination at a random speed. The random speed is distributed uniformly between zero (zero not included) and some maximum speed. We set the maximum speeds between 2 to 30 m/s for different scenarios. When the node reaches the destination, it again remains stationary for the *pause time* period and repeats the same procedure until the end of the simulation. We set the *pause time* to 20 seconds in our simulations which is good enough for a node to change the movement direction.

4.2.3. Communication Model

We allowed all the mobile nodes in the network to access the Internet, i.e., each mobile node sends data packets to the hosts in the Internet. Each mobile node in our simulation uses Constant Bit Rate (CBR) traffic to send packets to the corresponding hosts in the Internet. We wish to see the performance of different schemes under heavy traffic load. For this reason, we allow each mobile node to generate 10 packets per second and send them to the Internet. Like [4] [6] [9-12] [14-15] [19-21], we permit each mobile node in the MANET to generate packets of size 512 bytes. By varying the number of nodes, we actually varied the traffic load in different simulation scenarios.

4.2.4. Parameters

Table 1 gives the values of some simulation parameters that are used for most of the simulation scenarios.

Table 1. Common parameters used in most of the simulation scenarios.

Parameter	Value
Number of Internet gateways	4
Number of hosts in the Internet	2
Topology size	1000 x 1000 m^2
Transmission range	250 m
Internet BW	10 Mbps
Mobility Model	Random Waypoint
Traffic type	CBR
Packet size	512 bytes
Pause time	20 s
Simulation time	500 s

4.3. Result Analysis

Figures 4, 5 and 6 report IPDR, average end-to-end delay, and NCO respectively by varying the number of nodes but setting the maximum speed of a node to 30 m/s. In these figures we labelled our scheme as "interactive". We have taken the average of 10 simulation run results for each data point plotted in the figures.

When there are fewer nodes (less than 20) in the network, the total traffic generated by them is comparatively less. As a result, there is less congestion in the network which helps the nodes to deliver the packets to the gateways with less dropout and the gateways can also forward the packets to the Internet with ease. However, when the number of nodes in the network increases, the traffic load in the network also starts to increase.

Increased traffic load results in more congestion and more collisions in the network. As a result more packets are waiting in the interface queue of the forwarding nodes and getting dropped if the waiting time exceeds its limit. These facts reduce the packet delivery ratio and increase the end-to-end delay. Thus, IPDR decreases (Figure 4) and the average end-to-end delay increases (Figure 5) with the increase in the number of nodes in all the schemes. The periodic GWADV messages in the network in the other schemes help the nodes to have updated gateway information and achieve higher IPDR (Figure 4) with fewer nodes in the network.

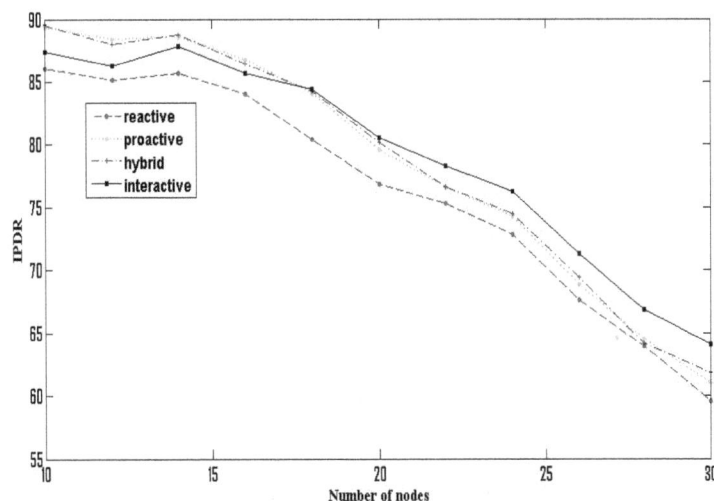

Figure 4. IPDR of all schemes against the number of nodes

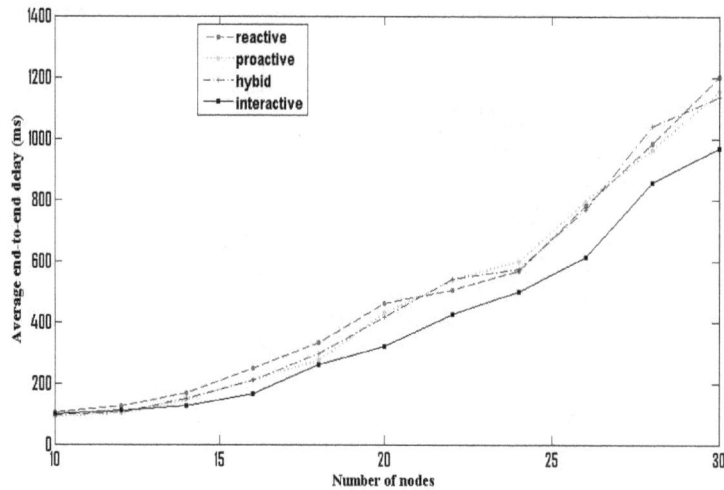

Figure 5. Average end-to-end delay of all schemes against the number of nodes

However, IPDR in our scheme started to exceed the IPDR of other schemes when the number of nodes is 20 or more. The average end-to-end delay obtained from our scheme is also better than that of other schemes (Figure 5). By avoiding the forwarding nodes having longer interface queue as well as the route to the gateway having higher concentration of neighbor nodes our scheme suffers from less packet drop and less waiting. For these reasons, IPDR is higher and the average end-to-end delay is lower in our scheme compared to that of other schemes while the number of nodes in MANET is increasing beyond 20.

From Figure 6 we can see that our scheme out performs the other schemes with respect to NCO performance metric. NCO obtained from all the schemes increase with the number of nodes in the network. Traffic load in the network increases as the number of nodes in MANET increases, which in turn increases the packet drop as explained earlier. Since NCO is the ratio between the number of routing packets and the number of successfully delivered data packets, it increases when there are less delivered data packets. As our scheme suffers from less packet drop than that of the others, it yields less NCO than that of others. Again, a gateway in our scheme broadcasts a GWADV message in response to a GWDSC message.

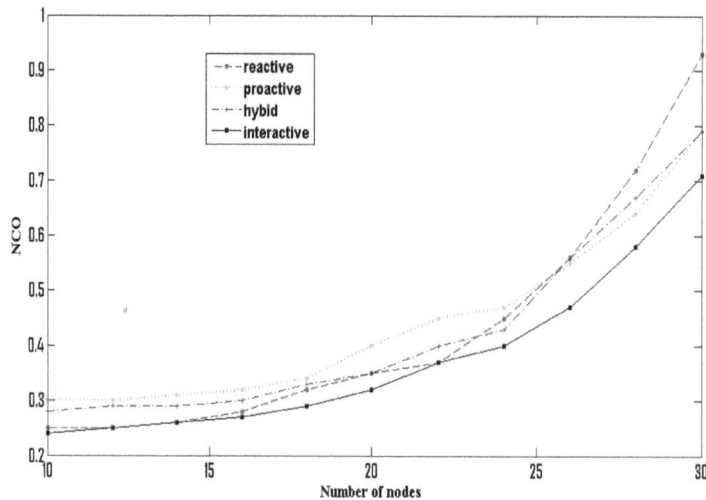

Figure 6. NCO of all schemes against the number of nodes

Not only the requesting mobile node gets the gateway information from the GWADV message but also the other nodes get the same information without transmitting their own GWDSC messages. This technique allows many mobile nodes to bypass the gateway discovery phase. As a result, they do not overwhelm the network by broadcasting GWDSC messages. For this reason, we have less routing packets in our scheme than that of other schemes, i.e., less NCO.

Figures 7, 8, and 9 report the same performance metrics respectively by varying the speed of the nodes but using only 30 mobile nodes. We have taken the average of 10 simulation run results for each data point plotted in the figures.

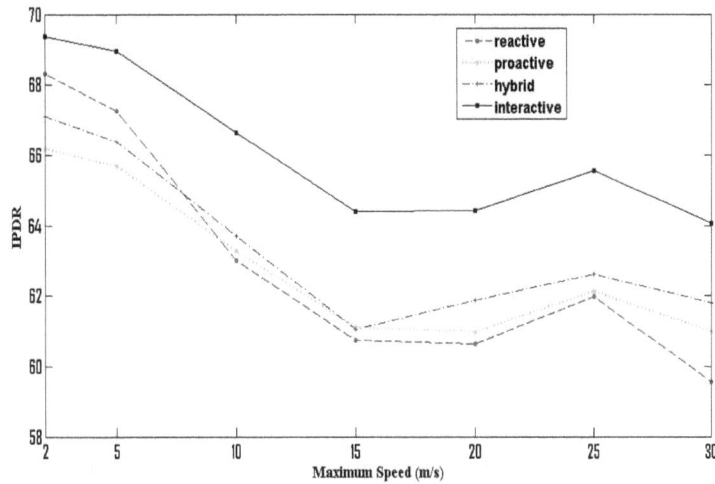

Figure 7. IPDR of all schemes against the speed of nodes

Figure 7 shows that IPDR obtained from all the schemes is high at the low speed, i.e. at 2m/s; it starts to decrease with the increase in the speed. The reason behind this fact is that the routing tables of the mobile nodes become obsolete when the nodes move with the high speed. As a result, more packets are dropped by the nodes in the network due to having no routes or obsolete routes to the gateways and the IPDR is reduced. Our scheme performs better than the other schemes by selecting gateways that have less dense route and the forwarding nodes on the route that have shorter queue lengths.

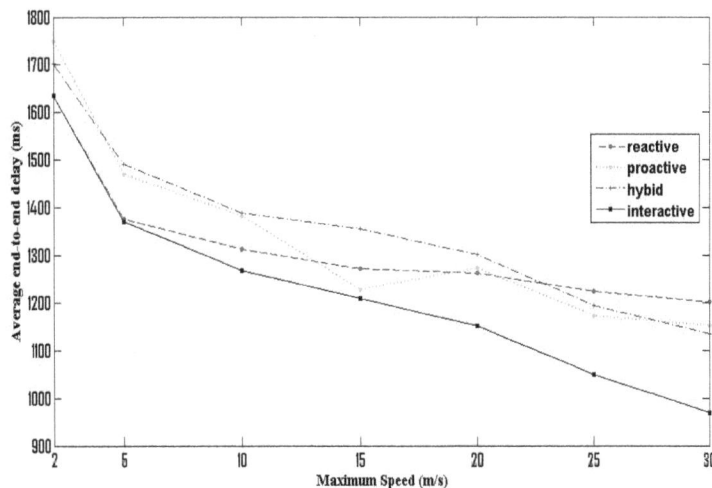

Figure 8. Average end-to-end delay of all schemes against the speed of nodes

We can see from Figure 8 that the average end-to-end delay in all the schemes decreases at the higher speeds. At the higher speeds the entries in the routing tables become obsolete quickly. Higher number of packets are dropped in the network for not having the routing entry. This reduces the average length of the interface queue in the network. Because of these shorter queue lengths, packets do not need to wait much in the network to get delivered. Our scheme avoids the routes having longer queue lengths and higher concentration of neighbor nodes. For this reason, our scheme experiences the lowest end-to-end delay.

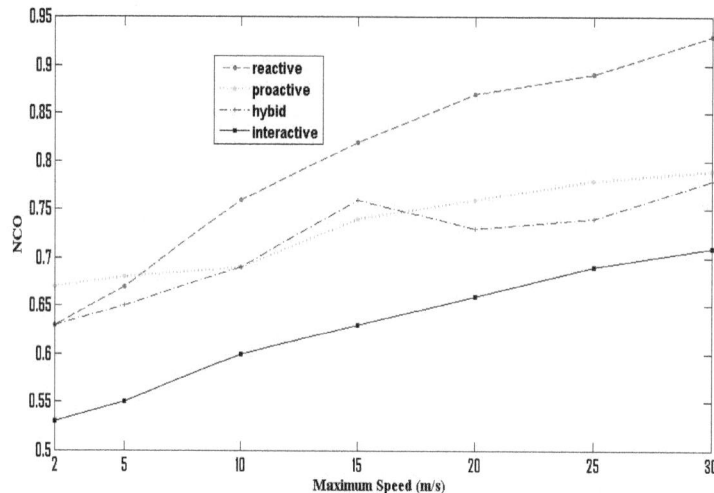

Figure 9. NCO of all schemes against the speed of nodes

Figure 9 shows that NCO, which is the ratio between the number of routing packets and the number of packets successfully delivered, increases with the speed of the mobile nodes in every scheme. Since the routing tables of the mobile nodes become obsolete when the nodes move with the high speed, nodes in the network suffer from having no routes or obsolete routes to the gateways. This fact causes more packet drops and more route re-discoveries. As a result NCO of all the schemes increases as the speed of the mobile nodes increases. However, our scheme has less NCO than that of the other schemes because it has less packet dropouts and it requires less routing packets compared to that of the other schemes.

From the above analysis of the results, we can conclude that our gateway discovery and selection scheme performs better than all other existing schemes in terms of packet delivery ratio, end-to-end delay, and network overhead with different size of MANET and with different speed of mobile nodes in the MANET. Thus, the proposed gateway discovery and selection scheme will scale well with the number of nodes, the traffic load and the speed of the nodes.

4.4. Statistical Analysis of the Simulation Results

We perform a statistical test to show that our scheme provides significant performance improvement over the other schemes.

We use the paired two-sample two-tailed t test to determine whether the improvement in the performance metrics i.e. IPDR, average end-to-end delay, and NCO in our scheme is significantly better than that of the reactive scheme. We compare two schemes in each data points given in the figures from Figure 5 to 10. Since each data point is the average of ten simulation run results, we simply measure the results of the reactive and interactive schemes in each run as the before and the after means respectively in order to get our t test results.

In our t test, the level of significance (**alpha**) is 0.05, the sample size (**n**) is equal to 10 and the degrees of freedom (**df**) is equal to (**n** – 1) = 9. The Critical t value ($T_{critical}$) for two tailed t test with **df** = 9 and **alpha** = 0.05 is 2.26215 [31]. We compare the t values (T_{value}) obtained from the t tests with the critical t value ($T_{critical}$) to determine whether there is a significant difference between the "reactive" and our "interactive" schemes. If a T_{value} is greater than the $T_{critical}$ then we reject the null hypothesis and if a T_{value} is smaller than the $T_{critical}$ then we accept the null hypothesis.

4.4.1. T test for IPDR

Table 2. T test results on IPDR in Figure 4.

Speed (m/s)	Schemes	Mean	Variance	T_{value}	T_{value} - $T_{critical}$	Remarks
2	reactive	68.313	35.14162	2.252716	-0.00944	accept H_0
	interactive	69.384	36.42932			
5	reactive	67.25	24.86528	3.899695	1.637538	reject H_0
	interactive	68.965	17.72547			
10	reactive	63.004	10.56805	6.589359	4.327202	reject H_0
	interactive	66.634	13.93272			
15	reactive	60.745	7.684339	14.07569	11.81353	reject H_0
	interactive	64.4	7.034222			
20	reactive	60.653	9.665534	8.291875	6.029718	reject H_0
	interactive	64.421	4.958988			
25	reactive	61.981	9.239654	8.397143	6.134986	reject H_0
	interactive	65.551	5.000868			
30	reactive	59.55	13.85924	10.7013	8.439143	reject H_0
	interactive	64.061	9.760157			

Table 3. T test results on IPDR in Figure 7.

No. of nodes	Scheme	Mean	Variance	T stat	T_{value} - $T_{critical}$	Remarks
10	reactive	86.075	24.62596	3.278258	1.016101	reject H_0
	interactive	87.402	18.21993			
12	reactive	85.181	8.546321	4.332188	2.070031	reject H_0
	interactive	86.304	6.662849			
14	reactive	85.692	12.84706	7.06506	4.802903	reject H_0
	interactive	87.891	13.06741			
16	reactive	84.01	5.872689	2.43274	0.170583	reject H_0
	interactive	85.701	8.154557			
18	reactive	80.374	12.1318	4.28	2.02	reject H_0
	interactive	84.404	10.37456			
20	reactive	76.882	10.8134	8.44477	6.18	reject H_0
	interactive	80.584	7.326204			
22	reactive	75.261	9.802143	7.111317	4.84916	reject H_0
	interactive	78.303	7.697712			
24	reactive	72.8	4.1636	13.83014	11.56798	reject H_0
	interactive	76.284	4.032316			
26	reactive	67.614	9.775161	5.8816	3.619443	reject H_0
	interactive	71.259	8.515699			
28	reactive	63.924	6.573827	6.61801	4.355853	reject H_0
	interactive	66.891	4.941062			
30	reactive	59.55	13.85924	10.7013	8.439143	reject H_0
	interactive	64.061	9.760157			

To perform the t test on IPDR of Figures 4 and 7, our null hypothesis and the alternate hypothesis are as follows:

H_0: The two means of the IPDR of the reactive and our interactive schemes are not significantly different.

H_a: The two means of the IPDR of the reactive and our interactive schemes are significantly different.

T test results on IPDR of Figures 4 and 7 are given in Tables 2 and 3 respectively. From Tables 2 and 3, we see that the difference between the T_{value} and the $T_{critical}$ is positive for most of the cases (we reject the null hypothesis), i.e., our interactive scheme provides higher IPDR than the reactive scheme for most of the cases with a confidence level 95%.

4.4.2. T test for Average end-to-end delay

Table 4. T test results on average end-to-end delay in Figure 5.

Speed (m/s)	Scheme	Mean	Variance	T stat	T_{value} - $T_{critical}$	Remarks
2	reactive	1635.244	111306.2	0.00015	-2.26201	accept H_0
	interactive	1635.79	102155.4			
5	reactive	1375.798	168763.5	0.073861	-2.1883	accept H_0
	interactive	1370.436	118051.5			
10	reactive	1312.123	59996.24	0.878159	-1.384	accept H_0
	interactive	1269.379	63131.87			
15	reactive	1271.642	33214.52	1.243227	-1.01893	accept H_0
	interactive	1209.519	43261.78			
20	reactive	1263.175	39371.22	3.049494	0.787337	reject H_0
	interactive	1151.85	28371.26			
25	reactive	1223.248	12716.83	3.932532	1.670375	reject H_0
	interactive	1050.294	24786.76			
30	reactive	1201.496	33353.4	8.879538	6.617381	reject H_0
	interactive	970.025	21784.99			

Table 5. T test results on average end-to-end delay in Figure 8.

No. of nodes	Scheme	Mean	Variance	T stat	T_{value} - $T_{critical}$	Remarks
10	reactive	106.402	905.7582	1.694366	-0.56779	accept H_0
	interactive	99.414	1294.542			
12	reactive	125.905	1023.625	3.23106	0.968903	reject H_0
	interactive	110.932	1015.462			
14	reactive	167.781	771.3989	7.235256	4.973099	reject H_0
	interactive	125.581	617.7584			
16	reactive	248.456	5355.004	4.078866	1.816709	reject H_0
	interactive	165.631	902.1428			
18	reactive	333.069	3229.179	4.859085	2.60	reject H_0
	interactive	260.976	3166.544			
20	reactive	461.234	4367.612	4.626608	2.36	reject H_0
	interactive	320.915	3509.737			
22	reactive	507.213	3056.937	2.085241	-0.17692	accept H_0
	interactive	426.488	9503.946			
24	reactive	568.126	2542.479	1.885491	-0.37667	accept H_0
	interactive	499.204	7242.416			
26	reactive	779.692	20805.49	5.547239	3.285082	reject H_0
	interactive	613.58	15840.75			
28	reactive	982.681	7900.833	3.393468	1.131311	reject H_0
	interactive	853.974	16227.58			
30	reactive	1201.496	33353.4	8.879538	6.617381	reject H_0
	interactive	970.025	21784.99			

To perform the t test on average end-to-end delay of Figures 5 and 8, our null hypothesis and the alternate hypothesis are as follows:

H_0: The two means of the delay of the reactive and our schemes are not significantly different.

H_a: The two means of the delay of the reactive and our schemes are significantly different.

T test results on delay of Figures 5 and 8 are given in Tables 4 and 5 respectively. From Tables 4 and 5, we see that our interactive scheme provides lower average end-to-end delay than the reactive scheme for most of the cases with a confidence level 95%.

4.4.3. T test for NCO

Table 6. T test results on NCO in Figure 6.

Speed (m/s)	Scheme	Mean	Variance	T stat	$T_{value} - T_{critical}$	Remarks
2	reactive	0.634	0.01785	4.03458	1.772423	reject H_0
	interactive	0.526	0.006582			
5	reactive	0.68	0.036778	4.24356	1.981403	reject H_0
	interactive	0.553	0.012779			
10	reactive	0.764	0.008649	7.38037	5.118213	reject H_0
	interactive	0.6	0.005756			
15	reactive	0.826	0.021404	5.95741	3.695253	reject H_0
	interactive	0.632	0.008951			
20	reactive	0.871	0.013877	7.81772	5.555563	reject H_0
	interactive	0.658	0.002929			
25	reactive	0.895	0.011072	7.70752	5.445363	reject H_0
	interactive	0.694	0.002761			
30	reactive	0.930	0.019662	6.30933	4.047173	reject H_0
	interactive	0.708	0.002529			

Table 7. T test results on NCO in Figure 9.

No. of nodes	Scheme	Mean	Variance	T stat	$T_{value} - T_{critical}$	Remarks
10	reactive	0.248	0.000262	2.44949	0.187333	reject H_0
	interactive	0.244	0.000227			
12	reactive	0.251	0.000143	0.317999	-1.94416	accept H_0
	interactive	0.252	0.000196			
14	reactive	0.261	0.000521	1.86052	-0.40164	accept H_0
	interactive	0.256	0.000293			
16	reactive	0.278	0.000596	2.75085	0.488693	reject H_0
	interactive	0.265	0.000428			
18	reactive	0.321	0.00061	6.81516	4.55E+00	reject H_0
	interactive	0.287	0.000357			
20	reactive	0.353	0.001312	4.30187	2.04E+00	reject H_0
	interactive	0.324	0.000804			
22	reactive	0.374	0.000671	0.58277	-1.67939	reject H_0
	interactive	0.37	0.000778			
24	reactive	0.449	0.001588	3.8512	1.589043	reject H_0
	interactive	0.403	0.001446			
26	reactive	0.559	0.004766	3.53363	1.271473	reject H_0
	interactive	0.474	0.003329			
28	reactive	0.724	0.007604	7.96496	5.702803	reject H_0
	interactive	0.583	0.002934			
30	reactive	0.930	0.019662	6.30933	4.047173	reject H_0
	interactive	0.708	0.002529			

To perform the t test on NCO of Figures 6 and 9, our null hypothesis and the alternate hypothesis are as follows:

H_0: The two means of the NCO of the reactive and our schemes are not significantly different.

H_a: The two means of the NCO of the reactive and our schemes are significantly different.

T test results on NCO of Figures 6 and 9 are given in Tables 6 and 7 respectively. From Tables 6 and 7 it is evident that our interactive scheme provides lower NCO than that of the reactive scheme for most of the cases with a confidence level 95%. All the t test results prove that our scheme is significantly better than the reactive scheme in terms of packet loss, end-to-end delay, and network overhead.

5. CONCLUSION

To rescue the network from the problems of current Internet gateway discovery and selection schemes, we presented a new gateway discovery and selection scheme. Our scheme uses a triggered broadcast of gateway advertisement messages at the gateways when being hit by gateway discovery messages. We also bounded the dissemination of the gateway advertisement messages up to the requesting mobile node from the gateway. We combined hop count, traffic load (interface queue length), and the total number of neighbors along a route to the gateway in order to formulate a new metric for gateway selection. Our metric chooses the gateway which is not only closest but also has the route from the mobile node with less load and less dense. We compared our gateway discovery and selection scheme with the other schemes in terms of three performance metrics: Internet Packet Delivery Ratio, Average End-to-End Delay and Normalized Control Overhead. Simulation results show that our scheme outperforms other schemes.

A number of open issues remain. In this research work, we consider the gateways to be stationary. In a hybrid environment, it is very likely that there will be a mixture of stationary and mobile gateways. Therefore, mobility of the gateways is an important issue in the gateway discovery and selection process and needs to be considered with due diligence. We considered much higher Internet bandwidth for the gateways compared to that of the MANET. However, higher Internet bandwidth might not be available at the gateways and it might be a serious bottleneck for the Internet traffic of the MANET. We allowed the gateway to broadcast gateway advertisement message when it is being hit by a gateway discovery message without considering the current traffic load at the gateway. If the current load is higher and new Internet traffic is directed towards this gateway by a gateway selection algorithm at the MANET nodes which does not consider the current traffic at the gateway, the new Internet traffic at the heavily loaded gateway might increase serious congestion in the network. In our future work, we will consider these open issues.

REFERENCES

[1]   Wakikawa R., Malinen J. T., Perkins C. E., Nilsson A., Tuominen A. J., "Global Connectivity for IPv6 Mobile Ad Hoc Networks," *IETF Internet Draft*, draft-wakikawa-manet-globalv6-05.txt, March 2006.

[2]   Sun Y., Belding-Royer E. M., Perkins C. E., "Internet Connectivity for Ad Hoc Mobile Networks," *International Journal of Wireless Information Networks, Special Issue on Mobile Ad Hoc Networks (MANETs): Standards, Research, Applications*, pp. 75-88, April 2002.

[3]   Jonsson U., Alriksson F., Larsson T., Johansson P., Maguire Jr. G. Q., "MIPMANET – Mobile IP for Mobile Ad Hoc Networks," *Proc. of the First IEEE/ACM Annual Workshop on Mobile Ad Hoc Networking and Computing*, USA, pp. 75-85, August 2000.

[4]   Kumar R., Misra M., Sarje A. K., "A Proactive Load-Aware Gateway Discovery in Ad Hoc Networks for Internet Connectivity," *International Journal of Computer Networks & Communications (IJCNC)*, vol. 2, no.5, pp. 120-139, September 2010.

[5]   Engelstad P. E., Tønnesen A., Hafslund A., Egeland G. "Internet Connectivity for Multi-Homed Proactive Ad Hoc Networks," *Proc. of the International Conference on Communication*, France, pp. 4050-4056, June 2004.

[6]   Khan K. U. R., Reddy A. V., Zaman R. U., Kumar M. "An Effective Gateway Discovery Mechanism in an Integrated Internet-MANET (IIM)," *Proc. of the International Conference on Advances in Computer Engineering*, India, pp. 24-28, June 2010.

[7]   Yuste A. J., Triviño A., Trujillo F. D., Casilari E., "Improved Scheme for Adaptive Gateway Discovery in Hybrid MANET," *Proc. of the International Conference on Distributed Computing Systems Workshops*, Genova, pp. 270-275, June 2010.

[8]   Kim Y., Ahn S., Yu H., Lee J., Lim Y., "Proactive Internet Gateway Discovery Mechanisms for Load-Balanced Internet Connectivity in MANET," *Proc. of International Conference on Information Networking. Towards Ubiquitous Networking and Services*, Portugal, pp. 285-294, January 2007.

[9]   Hamidian A. A., "A Study of Internet Connectivity for Mobile Ad Hoc Networks in NS 2," *Master's Thesis*, Department of Communication Systems, Lund Institute of Technology, Lund University, January 2003.

[10]   Le-Trung Q., Engelstad P. E., Skeie T., Taherkordi A., "Load-Balance of Intra/Inter-MANET Traffic over Multiple Internet Gateways," *Proc. of the International Conference on Advances in Mobile Computing and Multimedia*, Austria, pp. 50-57, November 2008.

[11]   Li X. , Li Z., "A MANET Accessing Internet Routing Algorithm based on Dynamic Gateway Adaptive Selection," *Frontiers of Computer Science in China*, vol. 4 , no. 1, pp. 143-150, March 2010.

[12]   Zhanyang X., Xiaoxuan H., Shunyi Z., "A Scheme of Multipath Gateway Discovery and Selection for MANET Using Multi-Metric," *Proc. of the International Conference on Information Science and Engineering*, China, pp. 2500-2503, December 2009.

[13]   Lee J., Kim D., Garcia-Luna-Aceves J. J., Choi Y., Choi J., Nam S., "Hybrid Gateway Advertisement Scheme for Connecting Mobile Ad Hoc Networks to the Internet," *Proc. of the 57th IEEE Vehicular Technology Conference*, Korea , pp. 191-195, April 2003.

[14]   Ratanchandani P., Kravets R., "A Hybrid Approach to Internet Connectivity for Mobile Ad Hoc Networks," *Proc. of IEEE Wireless Communications and Networking Conference*, USA, pp. 1522-1527, March 2003.

[15]   Ruiz P. M., Gomez-Skarmeta A. F., "Adaptive Gateway Discovery Mechanisms to Enhance Internet Connectivity for Mobile Ad Hoc Networks," *Ad Hoc and Sensor Wireless Networks*, vol. 1, pp. 159-177, March 2005.

[16]   Jiang H., Jin S., "Adaptive Strategies for Efficiently Locating Internet-based Servers in MANETs," *Proc. of ACM International Workshop on Modeling, Analysis, and Simulation of Wireless and Mobile Systems*, Canada, pp. 341-348, October 2005.

[17]   Bin S., Bingxin S., Bo L., Zhonggong H., Li Z., "Adaptive Gateway Discovery Scheme for Connecting Mobile Ad Hoc Networks to the Internet," *Proc. of International Conference on Wireless Communications, Networking and Mobile Computing*, China, vol. 2, pp. 795-799, September 2005.

[18]   Zhuang L., Liu Y., Liu K., Zhai L., Yang M., "An Adaptive Algorithm for Connecting Mobile Ad Hoc Network to Internet with Unidirectional Links Supported," *The Journal of China Universities of Posts and Telecommunications*, vol. 17, supplement 1, pp. 44-49, July 2010.

[19]   Park B., Lee W., Lee C., "QoS-aware Internet access schemes for wireless mobile ad hoc networks," *Computer Communications*, vol. 30, issue 2, pp. 369-384, January 2007.

[20]   Ma W., Liu J., "A Gateway Selection Scheme for Internetworking of MANET and Internet using Improved Genetic Algorithm," *Proceedings of the International Conference on Wireless Communications, Networking and Mobile Computing*, USA, pp. 2668-2671, September 2009.

[21]   Ruiz P. M., Ros F. J., Gomez-Skarmeta A. F., "Internet Connectivity for Mobile Ad Hoc Networks: Solutions and Challenges," *IEEE communication Magazine*, vol. 43, issue 10, pp. 118-125, October 2005.

[22]   Belding-Royer E. M., Sun Y., Perkins C. E., "Global Connectivity for IPv4 Mobile Ad Hoc Networks," *IETF Internet Draft*, draft-royer-manet-globalv4-00.txt, November 2001.

[23]   Geetha M., Umarani R., Kiruthika R., "A Comparative Study of Gateway Discovery Protocol in MANET," *International Journal of Computer Applications*, vol. 11, no. 2, pp. 16-22, December 2010.

[24]   Kassahun W., "Performance Evaluation and Nodes' Mobility Effect Analysis on AODV based Internet Gateway Discovery Algorithms in Social Networks," *Master's Thesis*, Department of Electrical and Computer Engineering, Faculty of Technology, School of Graduate Studies, Addis Ababa University, March 2010.

[25]   Perkins C. E., Das S. R., "Ad hoc On-Demand Distance Vector (AODV) Routing," *IETF Internet draft*, draft-ietf-manet-aodv-09.txt, November 2001.

[26]   Jacquet P., Muhlethaler P., Qayyum A., "Optimized Link State Routing Protocol," *IETF Internet Draft*, draft-ietf-manet-olsr-00.txt, November 1998.

[27]   Johnson D.B., Maltz D.A., "Dynamic Source Routing in Ad Hoc Wireless Networks," *In Mobile Computing, Kluwer Academic Publishers*, chapter 5, pp. 153-181, 1996.

[28]   NS-2 home page http://www.isi.edu/nsnam/ns/index.html.

[29]   AODV+: The Network Simulator: Contributed Code, available from: <http://www.isi.edu/nsnam/ns/ns-contributed.html/>.

[30]   Hyytiä E., Koskinen H., Lassila P., Penttinen A., Virtamo J., "Random Waypoint Model in Wireless Networks," *Networks and Algorithms: Complexity in physics and Computer Science* , Helsinki, June 2005.

[31]   Nehmzow U., "Statistical Tools for Describing Experimental Data," *Robot Behaviour: Design, Description, Analysis and Modelling*, Chapter 4, Springer, 2009.

9

PATH-CONSTRAINED DATA GATHERING SCHEME FOR WIRELESS SENSOR NETWORKS WITH MOBILE ELEMENTS

Bassam A. Alqaralleh and Khaled Almi'ani
Al-Hussein Bin Talal University, Jordan

ABSTRACT

Wireless Sensor Networks (WSNs) have emerged as a promising solution for variety of applications. Recently, in order to increase the lifetime of the network, many proposals have introduced the use of Mobile Elements (MEs) as a mechanical carrier to collect data. In this paper, we investigate the problem of designing the mobile element tour to visit subset of the nodes, termed as caching points, where the length of the mobile element tour is bounded by pre-determined length. Caching can be implemented at various points on the network such that any node in the network is at most k-hops away from one of these caching points. To address this problem, we present heuristic-based solution. Our solution works by partitioning the network such that the depth of each partition is bounded by k. Then, in each partition, the minimum number of required caching points is identified. We compare the resulting performance of our algorithm with the best known comparable schemes in the literature.

KEYWORDS

Wireless Sensor Networks, Mobile Element

1. INTRODUCTION

Wireless Sensor Networks (WSNs) composed of multifunctional miniature devices with sensing, computation and wireless communication capabilities. Such devices are normally battery operated. Thus, recharging a sensor node is impractical because they are typically designed for hostile environments; therefore, energy efficiency is critical. In typical static multi-hop data communication paradigm, the sensors around the sink are likely the first to run out of energy. This is due to the fact that these sensors carry heavier traffic loads. Once these sensor nodes fail, the operational lifetime of the networks ends and the network stops working because the entire network becomes unable to communicate with the sink. Therefore, reducing energy consumption in WSNs is becoming a major design challenge.

In order to significantly increase the lifetime of the network via considerably reducing the energy consumption, many proposals [1][2][3] have introduced the use of the Mobile Elements (MEs) as an efficient approach. A mobile element roams through the network and collects data from sensor nodes via short range (single-hop) communications. Therefore, in comparison with multi-hop communications the energy consumption can be considerably reduced. However, the speed of the mobile element is typically low[4][5], and therefore reducing the data gathering latency is important to ensure the efficiency of such approach.

To address this problem, several proposals presented a hybrid approach, which combines multi-hop forwarding with the use of mobile elements. In this approach, each mobile element visits subset of the nodes termed as caching points. These caching points store the data of the nodes that are not included in the tour of the mobile element. Once a mobile element becomes within the transmission range of a caching point, the caching point transmits its data to the mobile element. By adopting such an approach, the mobile element gathers the data of the entire network without the need for visiting each node physically.

In this paper, we focus on the problem of reducing the latency of the mobile elements tours. Accordingly, we investigate the K-Hop tour planning (KH-tour) problem. This problem deals with designing the shortest possible tour for the mobile, such that the tour start and end at the sink, and the maximum number of hops between any node and the tour is bounded by pre-determined value.

To address this problem, we present an algorithmic-based solution. This solution is based on partitioning the network into groups of partitions, where the depth of each partition is selected to ensure that one node in each partition can be identified as a caching point without violating the hop constraints. Then, the tour of the mobile element will be constructed to visit all nodes that have been identified as caching points.

The rest of the paper is organized as follows. Section 2 provides formal definition of the presented problem. Section 3 presents the related work in this area. Section 4 presents the heuristic solution of the presented problem. In Section 6, the evaluation for the proposed heuristics is presented. Finally, Section 7 concludes the paper.

2. RELATED WORKS

Many proposals in the recent literature have studied the use of mobile elements to reduce the energy consumption in sensor networks. In this section, we discuss the most related work in the literature.

The problem presented in our work can be recognized as an extended version of the tour scheduling problem presented by Somasundara et al. [5][6]. In this tour scheduling problem, the authors assume that each node must be visited within a time-deadline to avoid buffer overflow. Accordingly, the authors proposed different heuristics to determine the path of the mobile elements.

Guney et al. [7] formulated the problem of finding the optimal sink trajectory and the data flow routs as mixed integer programming formulations. Accordingly, they presented several heuristics to address this problem. Liang et al. [10]also presented a mixed integer programming formulation for similar problem, where they incorporate the constraint of the travelling to the formulation.

In [9], the authors proposed a data collection scheme aims to increase the network throughput. In this scheme, the mobile element is assumed to move periodically in pre-determined path, which is expected to visit a subset of the nodes called subsinks. The main goal of this scheme is optimizing the assignment of the sensor nodes to subsinks in order to increase the network throughput.

Zhao et al [11][12]investigated the problem of maximizing the overall network utility. Therefore, they presented two distributed algorithms for data gathering where the mobile sink stays at each anchor point (gathering point) for a period of sojourn time and collects data from nearby sensors via multi-hop communications. They considered both variable and fixed sojourn time.

The problem presented in this paper share some similarities with the Vehicle Routing Problem (VRP) [13]. Given a fleet of vehicles assigned to a depot, VRP deals with determining the fleet routes to deliver goods from a depot to customers while minimizing the vehicles' total travel cost.

Among the vehicle routing problem variations, the Vehicle Routing Problem with Time Windows (VRPTW) [14]is the closest to our problem. In VRPTW, each customer must be visited by exactly one vehicle and within a pre-defined time interval. The problem presented in this work also share some similarities with the Deadline Travelling Salesman Problem (Deadline-TSP)[14], which seeks the minimum tour length for a salesman to visit a set of cities, where each city must be visited before a pre-determined time deadline. In particular, the special case where all cites have the same deadline reduces Deadline-TSP to the well-known orienteering problem (OP) [15]; thus, our problem can be considered as a generalization of the orienteering problem.

The problem presented in this work shares some similarities with the problem proposed by Ma et al. [16]. In their proposal, they explored two settings where each polling point is either the location of a sensor or a point in the network area. On the other hand, the mobile element can communicate with one or more identified nodes in order to collect their data. Their goal is to design the mobile element tour which consists of polling points in order to find the shortest possible tour where each node is either on the tour or one-hop away from the tour.

The problem presented in this share some similarities with minimum-energy Rendezvous Planning Problem (RPP) [17][18]. In this problem, the objective is to determine the mobile element path such the total Euclidean distance between nodes not included in the tour and the tour is minimized. In [17], the authors presented the Rendezvous Design for Variable Tracks (RD-VT) algorithm. The process of this algorithm starts by construction the Steiner Minimum Tree (SMT) that connects the source nodes. Then, the obtained tree will be traversed in pre-order until no more nodes can be visited without violating the deadline constraint. In this algorithm the visited nodes is identified as the caching points. Xing et al. [3] provided a utility-based algorithm and address the optimal case for restricted version of the problem. Many proposals [19] have also investigated this problem.

3. PROBLEM DEFINITION

An instance of the KH-tour problem consists of an undirected complete graph $G = \langle V, E \rangle$, v_s. V is the set that represents the location of the sensor nodes in the network. v_s represents the location of the sink. E is the set of edges that represents the communication pattern in the network. For each pair of nodes $v_i, v_j \in V$ there is an edge between these two nodes, if they are within each other communication range. $d(v_i, v_j)$ represents the Euclidean distance that the mobile element will travel between nodes v_i and v_j. In addition, we use a pre-determined value k that represents the maximum number of hops allowed between any node and the tour.

A solution to the KH-tour problem consists of a tour (a path in G) that starts and ends in v_s, where the objective is to minimize the Euclidean distance of this tour, such that each node $v_i \in V$ is at most k-hops away from the tour.

4. INTEGER LINEAR PROGRAM FORMULATION

Let $V = \{v_0, v_1, ..., v_n\}$ be the set the represents the nodes in the network($v_s \in V$). Also, let $d(u, v)$ be the distance which the mobile element needs to travel to reach node v from node u. We use the $y_{u,v}$ as an (0-1) indicator for each pair of nodes $(u, v) \in V$ such that $y_{u,v} = 1$, if the mobile element travels between node v and node u, and 0 otherwise. Also, let z_u be an integer variable for each node $u \in V$, taking only positive values, showing the order in which the nodes are visited in the resulting tour. In addition, let $x_{u,v}$ be an integer variable for each pair of nodes $v, u \in V$ such that $x_{u,v} = 1$, if nodes v and u are one hop away from each other, and node u

stores node v data; 0 otherwise. Also, let $h(v, u)$ be the number of hops between nodes v and u. Given the Integer Linear Program for our problem can be given as follows:

Minimize

$$\sum_{i,j \in V} y_{i,j} \cdot d(i,j) \tag{1}$$

Subject to:

$$\sum_{i \in V} y_{j,i} \leq 1 \qquad \forall j \in V \tag{2}$$

$$\sum_{i \in V} y_{i,j} \leq 1 \qquad \forall j \in V \tag{3}$$

$$\sum_{i \in V} y_{i,j} - \sum_{i \in V} y_{j,i} = 0 \qquad \forall i \in V \tag{4}$$

$$x_{i,j} \cdot h(v_i, v_j) \leq k \qquad \forall i,j \in V \tag{5}$$

$$\sum_{i \in V} x_{i,j} + \sum_{i \in V} y_{j,i} = 1 \qquad \forall j \in V \tag{6}$$

$$z_i - z_j + n \cdot y_{i,j} \leq n - 1 \qquad \begin{array}{l} \forall j \in V/\{v_s\} \\ \forall i \in V \end{array} \tag{7}$$

Constraints (2-4) ensure that the load at each node is balanced. Constraint (5) ensures that the number of hops between any node and the tour is at most. Each node must be either involved in the mobile element tour or connected to a node involved in this tour, and this is represented by constraint (6). Constraint (7) is the sub-tour elimination constraint. Typically, it is hard to solve such problem. However, the ILP is given to take close look at the problem.

5. ALGORITHMIC-BASED SOLUTION

Our objective is to determine the shortest mobile element tour such that the number of hops between any node not included in the tour and the tour is at most k. Towards this end, in this section, we present the Graph Partitioning (GP) algorithm. The main idea of this algorithm is to divide the graph of the network into partitions, such that the depth of each partition is at most $2 \cdot k$ hops. As we will discuss in this section, this algorithm is designed to address the situation where the nodes are uniformly deployed, and the partitioning process aims to simplify the step of identifying the caching points.

The Graph Partitioning (GP) algorithm starts by identifying the center node of the graph. This node is the closest to all other nodes (in term of number of hops) inside the graph. Then the process iterates to identify the nodes belong to each partition. Once the nodes in each partition are identified, they are flagged by their partition number, to make sure that they will not be reconsidered during the partitioning steps. Then, the caching point identification step starts the process of identifying the caching points in each partition. This process aims to identify the minimum number of required caching points, such that the number of hops between any node and one of these nodes is at most k. As we will see next, this process works by establishing a path from the nodes, which have the highest number of neighbors. Two nodes are neighbors, if they are at most k-hops away from each other. Establishing this path aims to select the minimum number of required caching points. The last step of this algorithm is to construct the mobile element tour from the identified caching points. **Algorithm** 1 illustrates these steps. Now, we discuss the steps of this algorithm in more details.

Algorithm 1. The Graph Partitioning (GP) algorithm

1: **procedure**$GP(G, G', k)$

        Inputs: the graphs G and G'

k, maximum hops constraints

        Outputs: A tour (T) and forwarding trees (R)

2: $cn \leftarrow$ CenterNode(G)

3: $P \leftarrow$ Partition(G, cn)

4: $C \leftarrow$ CachingPoints(P, G)

5: $T \leftarrow$ TourBuild(C)

6: $R \leftarrow$ BuildRouting(T, G)

7: **procedure** CenterNode (G)

8:     **for** each vertex v in G**do**

9:       dist(v) $\leftarrow$ sum of hop-distances from all nodes

10:    **end for**

11:    **return** node with minimum dist(v)

12: **end procedure**

13: **procedure** Partition (G, cn)

14: **for** each vertex v in G**do**

15:    hops(v) $\leftarrow$ number of hops in the shortest path between v and cn

16:    $i \leftarrow \lfloor$hops(v)$/2 \cdot k\rfloor$+1

17:    **assign**v to P_i

18:   **end for**

19: **return**P

20:**end procedure**

21: **procedure** CachingPoints (G, P, k)

22:     **for** each P_i in P**do**

23:       **for** each v_i, v_j in P_i**do**

24:         **if** path between v_i and v_j at most k hops **do**

25:           **add** edge between v_i and v_j in G

26:         **end for**

27:     **end for**

28:     **for** each P_i in P**do**

29:       **add** node with highest degree to C_i

31:       **Remove** this node and its one-hope neighbors from P_i

32:     **end for**

33:     **return**P

34:**end procedure**

5.1 The partitioning step

This step starts by identifying the center node of the graph. Then, starting by the first partition, the process iterate to identify the nodes belong to each partition. The identity of the nodes that are assigned to each partition is selected based on the partition number, and the number of hops between any node and center node. For partition i, the set of the nodes that belong to this partition consists of the nodes that are not assigned to any previous partition, and can reach the center node within $i \cdot 2 \cdot k$ hops. As we can see, by performing such partitioning process, the network will be divided into partitions, where the depth of each partition is $2 \cdot k$. Now, in each partition there must be a set of nodes that can reach all other nodes inside the partition within k-hops. By performing this partitioning step, we attempt to reduce the solution space for selecting the caching points in an efficient manner in order to simplify the process of identifying the caching nodes.

5.2 The caching point identification step

In this step, the process iterates to determine the caching points for each partition. Starting at partition number1, for each pair of nodes $v_i, v_j \in V$ belong to this partition, we add an edge between these two nodes to E, if they are k-hops away from each other. Only the nodes that belong to the same partition will be considered during the process of determining the k-hops connectivity. Once the k-hops connectivity for all partitioned are examined, the process proceed by constructing a set of the caching points for each partition $C = [C_1, C_2, ..., C_i]$. In each partition(i), based on the degree of the nodes, subset of the partition nodes will be added to C_i. In each iteration, the node with the highest degree will be add to C_i. This node and all of its one-hop neighbors (inside the partition) will be flagged to ensure that they are considered in consecutive iterations. This process stops when all nodes inside each partition are flagged. Now, the set of the caching points consists of the collection of caching points in all partitions.

5.3 The routing trees construction and tour building steps

Once the caching points set are identified, each node not included in this set will be assigned to its nearest caching point. Then, for each caching point and the nodes which are assigned to this caching point, a Minimum Spanning Tree (MST) is created to establish multi-hop forwarding trees. The tour of the mobile element consists of the caching points and the sink. This is established using Christofides algorithm [18].

6. EXPERIMENTAL EVALUATION

To evaluate the presented algorithm's performance, we conducted an extensive set of experiments using the J-sim simulator [19]. the area of the network is 250,000m^2, and The radio parameters are set according to the MICAz data sheet [20], namely: the radio bandwidth is 250 Kbps, the transmission power is 21 mW, the receiving power is 15 mW, and the initial battery power is 10 Joules. The packet has a fixed size of 100 bytes. Each experiment is taken over an average of 10 different realizations of random topologies. We are particularly interested in investigating the following metrics:

- The lifetime of the network
- Number of caching points.

The parameters we consider in our experiments look at varying the number of nodes. We consider the following deployment scenarios:

- Uniform density deployment: in this scenario, we assume that the nodes are uniformly deployed in a square area of 500× 500m^2.
- Variable density deployment: in this scenario, we divide the network into a 10×10 grid of squares, where each square is 50×50m^2. We randomly choose 30 of the squares, and in each one of those squares we fix the node density to be x times the density in the remaining squares. x is a density parameter, which in most experiments (unless mentioned otherwise) is set to $x = 5$.

To benchmark the presented algorithm performance, we compared it against the Spanning Tree Covering Algorithm, which is proposed by Ma et al.[16], we will refer to this algorithm as the T-Covering algorithm. This algorithm starts by selecting the sink as the initial point in the mobile element tour. Then, in each iteration and based on a cost function, the polling point with the lowest cost will be added to the tour. Ma et al.[18] defines the polling point as a point in the network, where the mobile element can communicate with one or more sensor nodes via a single hop transmission. For a given polling point, this cost function returns the shortest distance between this polling point and a node belong to the tour, divided by the number of new nodes will be covered if this polling point is selected. This process iterate until all nodes are covered by the tour. To ensure the fairness of the comparison, we restrict the

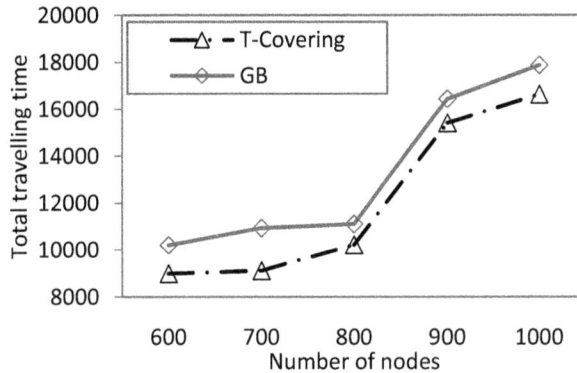

Figure 1: Total travelling time against the number of nodes, for the uniform density deployment scenario

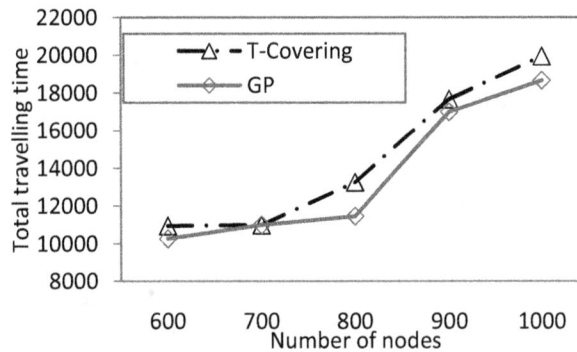

Figure 2: Total travelling time against the number of nodes, for the variable density deployment scenario

polling points in the T-covering algorithm to be the actual sensor nodes, and in the T-covering algorithm, we add a direct edge between any two nodes, if they are k-hops from each other.

First, we evaluate the impact of the number of nodes on the number of tours each algorithm obtains. Figure 1 and Figure 2 show the results for deployment scenarios. From the figures, we can see that in the uniform deployment scenario, the T-Covering algorithm outperforms the GP algorithm, where in the variable deployment scenario, the GP algorithm outperform the T-Covering algorithm. To understand this behavior, let us clarify the relationship between the deployment scenario and the expected performance for each algorithm. In the T-Covering algorithm, the employed cost function selects the node with the lowest cost to be included in the tour. The cost value for each node is obtained by dividing the distance between this node and the tour, by the number of nodes which will be covered if this node is included in the tour. Such cost function results in degrading

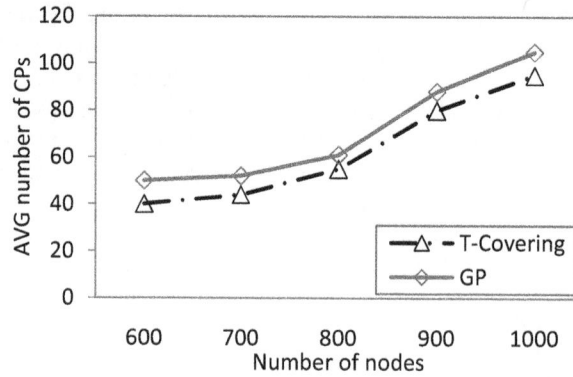

Figure 3: Average number of caching points against the number of nodes, for the uniform density deployment scenario

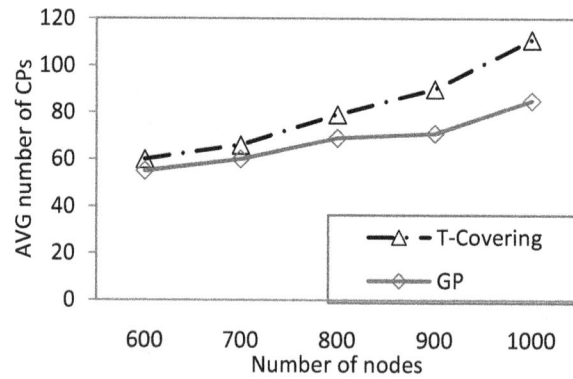

Figure 4: Average number of caching points against the number of nodes, for the variable density deployment scenario

the performance of the T-Covering algorithm in the variable deployment scenario compared to the uniform deployment scenario. This occurs because in the uniform deployment scenario, we expect that the T-Covering algorithm will select caching points uniformly located across the entire network. Furthermore, in the variable deployment scenario, this cost function is expected to result in selecting caching points mainly in the dense areas in the network and this is expected to increase the length of the obtained tour. In the GP algorithm, the process of partitioning the network results in disconnecting the process of selecting the caching points from the deployment distribution. These are the main factors which explain the performance of our algorithm.

We proceed to investigate the impact of the number of nodes on the number of caching points that each algorithm obtains. Figures 3 and 4 show the results for both the uniform density and the variable density deployment scenarios; respectively. From the figures, we can see that in the uniformly deployment scenario, the T-Covering algorithm obtains slightly lower number of caching points compared to the GP algorithm. On the other hand, in variable deployment scenario, the GP algorithm obtains significantly lower number of caching points. This is also due to the factors mentioned above, since in the variable deployment scenario, we expect that the T-covering algorithm obtains longer tour compared to the GP algorithm.

7. CONCLUSIONS

In this work, we address the problem of designing the mobile element tour, such that the length of this tour is below a pre-determined value, and each node is at most k-hops away from the tour. Accordingly, we propose a heuristic-based solution which is based on dividing the network into partitions, with pre-determined depth. Then, a subset of the nodes in each partition which will act as caching points for data must be identified such that the length of the tour that consists of these caching points and the sink node is below L.

An interesting open problem is considering some application scenarios where the latency requirements of data gathering may vary from one network's location to another. For example, some areas in the network need to be visited more frequently than others. In this case the tour length constraints would be different for different areas.

REFERENCES

[1] Y. Gu, D. Bozdag, E. Ekici, F. Ozguner, and C. G. Lee, "Partitioning based mobile element scheduling in wireless sensor networks," in Sensor and Ad Hoc Communications and Networks, 2005. IEEE SECON 2005. 2005 Second Annual IEEE Communications Society Conference on, 2005, pp. 386–395.

[2] G. Xing, T. Wang, Z. Xie, and W. Jia, "Rendezvous planning in wireless sensor networks with mobile elements," IEEE Trans. Mob. Comput., vol. 7, no. 12, pp. 1430–1443, Dec. 2008.

[3] K. Dantu, M. Rahimi, H. Shah, S. Babel, A. Dhariwal, and G. S. Sukhatme, "Robomote: enabling mobility in sensor networks," in Proceedings of the 4th international symposium on Information processing in sensor networks (IPSN), 2005, pp. 404 – 409.

[4] R. Pon, M. A. Batalin, J. Gordon, A. Kansal, D. Liu, M. Rahimi, L. Shirachi, Y. Yu, M. Hansen, W. J. Kaiser, and others, "Networked infomechanical systems: a mobile embedded networked sensor platform," in Proceedings of the 4th international symposium on Information processing in sensor networks, 2005, pp. 376 – 381.

[5] A. Somasundara, A. Ramamoorthy, and B. Srivastava, "Mobile Element Scheduling for Efficient Data Collection in Wireless Sensor Networks with Dynamic Deadlines," in Real-Time Systems Symposium, 2004. Proceedings. 25th IEEE International, 2004, pp. 296 – 305.

[6] A. A. Somasundara, A. Ramamoorthy, and M. B. Srivastava, "Mobile element scheduling with dynamic deadlines," IEEE Trans. Mob. Comput., vol. 6, no. 4, pp. 395–410, 2007.

[7] E. Güney, I. K. Altmel, N. Aras, and C. Ersoy, "Efficient integer programming formulations for optimum sink location and routing in wireless sensor networks," in Proceedings of the 23rd International Symposium on Computer and Information Sciences, 2008, pp. 1–6.

[8] W. Liang, J. Luo, and X. Xu, "Prolonging Network Lifetime via a Controlled Mobile Sink in Wireless Sensor Networks," in Proceedings of Global Telecommunications Conference (GLOBECOM 2010), 2010, pp. 1–6.

[9] B. Hull, V. Bychkovsky, Y. Zhang, K. Chen, M. Goraczko, A. Miu, E. Shih, H. Balakrishnan, and S. Madden, "CarTel: a distributed mobile sensor computing system," in Proceedings of the 4th international conference on Embedded networked sensor systems, 2006, pp. 1–14.

[10] M. Zhao and Y. Yang, "Optimization-Based Distributed Algorithms for Mobile Data Gathering in Wireless Sensor Networks," IEEE Trans. Mob. Comput., vol. 11, no. 10, pp. 1464–1477, 2012.

[11] M. Zhao and Y. Yang, "efficient data gathering with mobile collectors and space-division multiple access technique in wireless sensor networks," IEEE Trans. Comput., vol. 60, no. 3, pp. 400–417, 2011.

[12] P. Toth and D. Vigo, "The Vehicle Routing Problem," Soc. Ind. Appl. Math., 2001.

[13] M. M. Solomon, "ALGORITHMS FOR AND SCHEDULING PROBLEMS THE VEHICLE ROUTING WITH TIME WINDOW CONSTRAINTSS," Oper. Res., vol. 35, no. 2, pp. 254–265, 2010.

[14] B. Golden, L. Levy, and R. Vohra, "The orienteering problem," Nav. Res. Logist., vol. 34, no. 3, pp. 307–318, 2006.

[15] M. Ma, Y. Yang, and M. Zhao, "Tour Planning for Mobile Data-Gathering Mechanisms in Wireless Sensor Networks," IEEE Trans. Veh. Technol., 2013.

[16] G. Xing, T. Wang, W. Jia, and M. Li, "Rendezvous design algorithms for wireless sensor networks with a mobile base station," in Proceedings of the 9th ACM international symposium on Mobile ad hoc networking and computing (MobiHoc), 2008, pp. 231–240.

[17] G. Xing, T. Wang, Z. Xie, and W. Jia, "Rendezvous planning in mobility-assisted wireless sensor networks," in proceedings of the 28th IEEE InternationalReal-Time Systems Symposium (RTSS), 2007, pp. 311 – 320.

[18] N. Christofides, "Worst-case analysis of a new heuristic for the traveling salesman problem," 1976.

[19] J. C. Hou, L. Kung, N. Li, H. Zhang, W. Chen, H. Tyan, and H. Lim, "J-Sim: A Simulation and emulation environment for wireless sensor networks," IEEE Wirel. Commun. Mag., vol. 13, no. 4, pp. 104–119, 2006.

[20] J. L. Hill and D. E. Culler, "Mica: A wireless platform for deeply embedded networks," IEEE micro, vol. 22, no. 6, pp. 12–24, 2002.

10

COMPARING THE IMPACT OF MOBILE NODES ARRIVAL PATTERNS IN MANETS USING POISSON AND PARETO MODELS

John Tengviel[1], and K. Diawuo[2]

[1]Department of Computer Science, Sunyani Polytechnic, Sunyani, Ghana
[2]Department of Computer Engineering, KNUST, Kumasi, Ghana

ABSTRACT

Mobile Ad hoc Networks (MANETs) are dynamic networks populated by mobile stations, or mobile nodes (MNs). Mobility model is a hot topic in many areas, for example, protocol evaluation, network performance analysis and so on.How to simulate MNs mobility is the problem we should consider if we want to build an accurate mobility model. When new nodes can join and other nodes can leave the network and therefore the topology is dynamic.Specifically, MANETs consist of a collection of nodes randomly placed in a line (not necessarily straight). MANETs do appear in many real-world network applications such as a vehicular MANETs built along a highway in a city environment or people in a particular location. MNs in MANETs are usually laptops, PDAs or mobile phones.

This paper presents comparative results that have been carried out via Matlab software simulation. The study investigates the impact of mobility predictive models on mobile nodes' parameters such as, the arrival rate and the size of mobile nodes in a given area using Pareto and Poisson distributions. The results have indicated that mobile nodes' arrival rates may have influence on MNs population (as a larger number) in a location. The Pareto distribution is more reflective of the modeling mobility for MANETs than the Poisson distribution.

KEYWORDS

Mobility Models, MANETs, Mobile Nodes Distribution, Arrival Patterns, Pareto Distribution, Poisson Distribution, Matlab Simulation.

1. INTRODUCTION

Mobile Ad-hoc NETworks (*MANETs*) is a collection of wireless mobile nodes configured to communicate amongst each other without the aid of an existing infrastructure. MANETS are *Multi-Hop* wireless networks since one node may not be indirect communication range of other node. In such cases the data from the original sender has to travel a number of hops (hop is one communication link) in order to reach the destination. The intermediate nodes act as routers and forward the data packets till the destination is reached [1].

Recently, with the deployment of all kinds of wireless devices, wireless communication is becoming more important. In this research area, Ad-Hoc network is a hot topic which has attracted much of research attentions. A wireless ad hoc network is a decentralized wireless network. The network is ad hoc because it does not rely on a preexisting infrastructure, such as routers in wired networks or access points in managed (infrastructure) wireless networks. Instead, each node participates in routing by forwarding data for other nodes, and so the determination of which nodes forward data is made dynamically based on the network connectivity [2]. There are different kinds of routing protocol defined by how messages are sent from the source node to the destination node.

Based on this, it's reasonable to consider node mobility as an essential topic of ad-hoc network. With an accurate mobility model which represents nodes movement, designers can evaluate performance of protocols, predict user distribution, plan network resources allocation and so on. It can also be used in healthcare or traffic control area rescue mission, and so on.

Ad hoc networks are viewed to be suitable for all situations in which a temporary communication is desired. The technology was initially developed keeping in mind the military applications [3] such as battle field in an unknown territory where an infrastructure network is almost impossible to have or maintain. In such situations, the ad hoc networks having self-organizing [4] capability can be effectively used where other technologies either fail or cannot be effectively deployed. The entire network is mobile, and the individual terminals are allowed to move freely. Since, the nodes are mobile; the network topology is thus dynamic. This leads to frequent and unpredictable connectivity changes. In this dynamic topology, some pairs of terminals may not be able to communicate directly with each other and have to rely on some other terminals so that the messages are been delivered to their destinations. Such networks are often referred to as multi-hops or store-and-forward networks [5].

This paper presents a study on mobile nodes arrival patterns in MANETs using Poisson and Pareto models. Though not very realistic from a practical point of view, a model based on the exponential distribution can be of great importance to provide an insight into the mobile nodes arrival pattern. The section 2 illustrates a brief review on MANETs studies. The section 3 introduces the Poisson and Pareto distribution models. The simulation procedures and considered parameters are presented in section 4. The obtained results are objects in section 5 and the section 6 closes the paper to further research works.

2. RELATED WORKS

Currently there are two types [6, 7]of mobility models used in simulation of networks. These are traces and synthetic models. Traces are those mobility patterns that are observed in real-life systems. Traces provide accurate information, especially when they involve a large number of mobile nodes (MNs) and appropriate long observation period. On the other hand, synthetic models attempt to realistically represent the behaviour of MNs without the use of traces. They are divided into two categories, entity mobility models and group mobility models [1, 8, 9]. The entity mobility models randomise the movements of each individual node and represent MNs whose movements are independent of each other. However, the group mobility models are a set of groups' nodes that stay close to each other and then randomise the movements of the group and represent MNs whose movements are dependent on each other. The node positions may also vary randomly around the group reference point. In [10], the mobility study in ad hoc has been approximated to pedestrian in the street, willing to exchange content (multimedia files, mp3, etc.) with their handset whilst walking at a relative low speed. Some researchers have proposed basic mobility models such as Random Walk, Random Waypoint, [3, 4], etc. for performance comparison of various routing protocols. The concern with these basic designed models is that they represent a specific scenarios not often found in real lives. Hence their use in ad hoc network studies is very limited. Random Walk or Random Waypoint model though simple and elegant, produce random source of entry into a location with scattered pattern around the simulation area, sudden stops and sharp turns. In real-life, this may not really be the case.

3. MODELS OF STUDY

3.1. POISSON ARRIVAL DISTRIBUTION (NUMBER OF NODES)

When arrivals occur at random, the information of interest is the probability of n arrivals in a given time period, where $n = 0, 1, 2, \ldots\ldots$ n-1

Let λ be a constant representing the average rate of arrival of nodes and consider a small time interval Δt, with $\Delta t \rightarrow 0$. The assumptions for this process are as follows:

- The probability of one arrival in an interval of Δt seconds, say **(t, t+Δt)** is $\lambda\Delta t$, independent of arrivals in any time interval not overlapping **(t, t+Δt).**

- The probability of no arrivals in Δt seconds is **1-$\lambda\Delta$t**, under such conditions, it can be shown that the probability of exactly **n** nodes arriving during an interval of length of **t** is given by the Poisson distribution law [11] in equation 1:

$$P(n) = \frac{(\lambda t)^n e^{-\lambda t}}{n!}, \quad \text{where} \, n \geq 0, \, t > 0. \qquad (1)$$

The assumption of Poisson MN arrivals also implies a distribution of the time intervals between the arrivals of successive MN in a location.

3.2. PARETO DISTRIBUTION

The Pareto distributions [12-14] are characterized by two parameters: α and β. Parameter α is called shape parameter that determines heavy-tailed characteristics and $\beta = 1$ is called cutoff or the location parameter that determines the average of inter-arrival time.

The node arrival times of the Pareto distribution are independent and identically distributed, which means that each arrival time has the same probability distribution as the other arrival times and all are mutually independent. The two main parameters of the Pareto process are the shape α and the scale parameter (x).

For one parameter Pareto (α shape only), the distribution function can be written as equation 2:

$$F(X) = 1 - \left(\frac{1}{1+X}\right)^\alpha, X \geq 0 \qquad (2)$$

The pdf is given as in equation 3:

$$f(X) = \frac{u}{(1+X)^{\alpha+1}} \qquad (3)$$

and for the two – parameter Pareto distribution function defined over the real numbers can be written as in (4):

$$\begin{cases} F(X) = 1 - \left(\frac{1}{\alpha+X}\right)^{\beta} \\ , \\ X \geq 0; \quad \alpha, \beta > 0 \end{cases} \tag{4}$$

Its pdf is given as in equation 5:

$$f(X) = \frac{\alpha}{\beta} * \left(\frac{\beta}{X}\right)^{\alpha} \tag{5}$$

4. METHODOLOGY

4.1. Varying of α in Pareto Arrival Distribution

We assume the arrival distribution on the MNs population by using Pareto distributions.

Table 1: Varying α parameter values

Scenario	1	2	3	4	5
α (B)	0.3	0.4	0.5	0.8	0.9

For the simulations purposes, the varying α values are been considered. Heavy-tail is been modeled by a Pareto distribution and the main principle can be attributed to the principle of number of nodes. We have performed the simulations for a wide range of parameter values as in Table 1 for both one-parameter and two-parameter Pareto models.

4.2. VARYING OF ARRIVAL RATES FOR NODE DISTRIBUTION

The arrival pattern of mobile nodes has an impact on the performance of the network. In this scope, we have decided to analysis the effect of arrival distribution on the MNs population in a given area by using Poisson distribution as in equation 1.In most real-world MANETs, the node population in an area of interest varies with time. In this simulation, it is therefore necessary to investigate the impact of arrivals of MNs on the MANETs mobility.

The simulation area does not change as the arrival rate changes. The different values of arrival rates being considered in this study are shown in Table 2.

Table 2: Varying Arrival Rates

Scenario	1	2	3	4	5
Arrival rates	0.3	0.4	0.5	0.8	0.9

During the simulation, nodes were allowed to enter the location from a common source (0 degrees) but not from different sources. The number of MNs that entered the location was assumed to be Poisson distributed with varying arrival rates.

5. RESULTS AND DISCUSSION

5.1. Comparative Study using Pareto Arrival Pattern

In this section, the effect of arrival rates on MNs distribution and population in a defined location is analyzed as shown in Figure 1. It was observed that the various arrival rates increased the number of MNs also increased but to a certain limit. It is therefore the indication that every location has a limit or capacity of MNs it can contain.

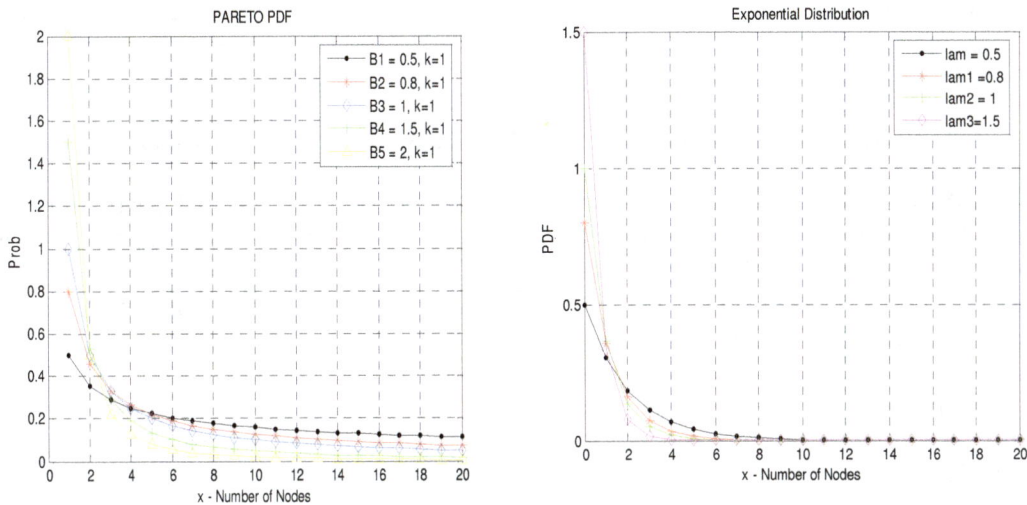

Figure 1: Single Parameter for Varying Values for B, and Exponential for Twenty Nodes

Figure 1may indicate that the exponential distribution was higher than the single parameter in the initial stages but as time progresses the exponential decreases fast to zero. The single parameter Pareto overtakes the exponential as the number of nodes increases and indication that the single parameter performs better than exponential distribution.

The Pareto distribution may show tail that decays much more slowly than the exponential distribution. The alpha is the shape parameter which determines the characteristics "decay" of the distribution (tail index) and A is the location parameter which defines the minimum value of x (number of nodes).

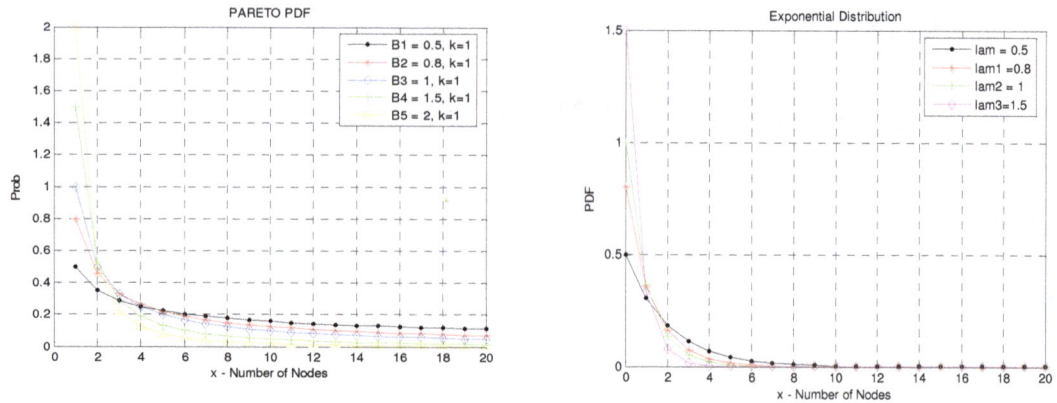

Figure 2: Two Parameter Pareto for Varying B Values and Exponential

In Figure2 the comparison between the two-parameter Pareto and exponential distributions is illustrated. It is obvious that the two-parameter Pareto outweighs the exponential distribution as the number of MNs increases. The exponential distributions decays very fast and finally get to the a-axis unlike the two-parameter Pareto distribution where some of the arrival rates distribution has not decay to zero.

However the two-parameter Pareto performed well than the one-parameter Pareto, since some of the arrival of the two-parameter did not decay to zero. The long-tailed nature of the two-parameter Pareto helped to clear out any congestion in a location when the arrival rate was small and the reverse was also true.

5.2 Effect of Varying Arrival Rates

In Figure 3, the effect of varying nodes' arrival rate is computed using Poisson model. Nodes may arrive at a location either in some regular pattern or in a totally random fashion. The arrival rates have shown to impact on the number of nodes in a particular location, although every location has a limited capacity. A high number of nodes typically translate into a higher average number of neighbours per node, which influences the route availability.

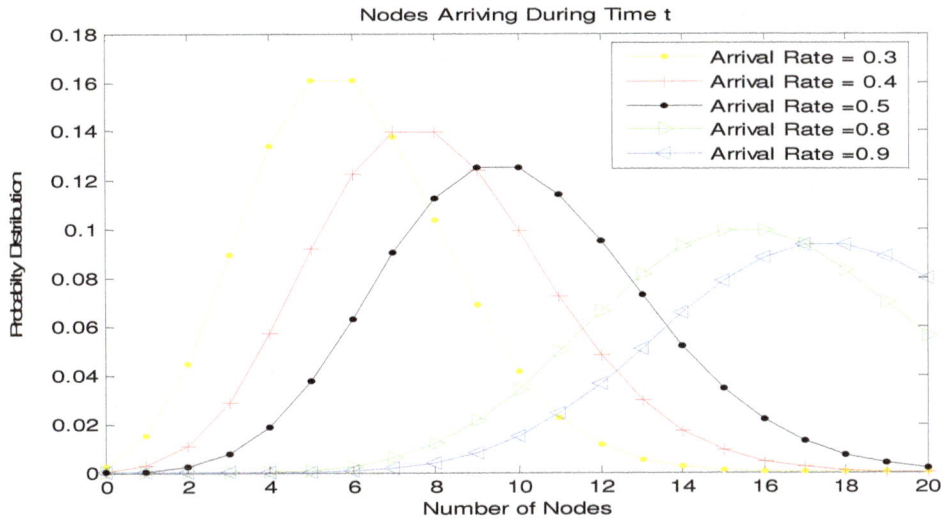

Figure 3: For Twenty Number of Nodes for varying Arrival rates

In reality, the total connection time of a node over a specific interval depends on the nodes encounter rate and the time in each encounter, both of which depend on the relative mobility of nodes.

Although a high node arrivals results in more node encounters, the network would eventually become congested. The impact of this relationship is that nodes can and will be tightly packed (i.e. High density) if their arrival rates is high (congestion), but if the arrivals is lower, the nodes must be farther apart (low density). For instance it is clear that there is some congestion for arrivals of MNs, since they have to follow some holding paths.

As the value of arrival rate increases, the shape of the distribution changes dramatically to a more symmetrical ("normal") form and the probability of a larger number of arrivals increases with increasing number of MNs. An interesting observation is that as the arrival rate increases, the properties of the Poisson distribution approach those of the normal distribution as in Figure3.

The first arrival processes of nodes give higher contact probabilities at higher arriving rates. This is due to the nodes' contiguity one to another making mobility difficult. In practice, one may record the actual number of arrivals over a period and then compare the frequency of distribution of the observed number of arrival to the Poisson distribution to investigate its approximation of the arrival distribution.

6. CONCLUSION

The arrival patterns have shown some impact on the network population, as the arrival rate increases the MNs population also increases to a peak and then decays rapidly to the x-axis. It was realized that the Poisson distribution is not good for the arrival distribution; therefore the Pareto distribution was considered. It has come out clear that the Pareto distribution is good for the arrival distribution, especially the two-parameter Pareto distribution which performed better than the single Pareto and exponential distributions even though at the earlier stages the exponential performed than the single Pareto distribution with a faster decay.

It may subsequently be admitted that mobility in MANETs is a difficult work and actually. It is an interesting research area that has been growing in recent years. Its difficulty is mainly generated because of the continuous changes in the network topology with time. The topological changes have impact on mobility techniques developed for infrastructure-based networks thus may not be directly applied to mobile adhoc networks. We have investigated through simulation mobility prediction of MNs using the queueing model.

REFERENCES

[1]   J. Boleng, T. Camp, and V. Tolety. "A Suvey of Mobility Models for Ad hoc Network Research", In Wireless Communication and Mobile Computing (WCMC), Vol. 2, No. 5, pages 483 – 502, 2002.

[2]   Subir Kumar Sarkar, T. G. Basavaraju, C. Puttamadappa, "Mobility Models for Mobile Ad Hoc Networks", 2007, Page 267 – 277, Auerbach Publications – www.auerbach-publications.comVolume 2, No.5, 2002.

[3]   C. Rajabhushanam and A. Kathirvel, "Survey of Wireless MANET Application in Battlefield Operations", (IJACSA) International Journal of Advanced Computer Science and Applications, Vol. 2, No.1, January 2011.

[4]   Buttyan L., and Hubaux J. P., "Stimulating cooperation in self-organizing mobile ad hoc networks. Mobile Networks andApplications: Special Issue on Mobile Ad Hoc Networks, 8(5), 2003.

[5]   C.P.Agrawal, O.P.Vyas and M.K Tiwari, "Evaluation of Varying Mobility Models & Network Loads on DSDV Protocol of MANETs", International Journal on Computer Science and Engineering Vol.1 (2), 2009, pp. 40 - 46.

[6]   P. N. Pathirana, A. V. Savkin& S. K. Jha. "Mobility modeling and trajectory prediction for cellular networks with mobile base stations". MobiHoc 2003: 213 -221.

[7]   MohdIzuanMohdSaad and Zuriati Ahmad Zukarnain, "Performance Analysis of Random-Based Mobility Models in MANET Routing Protocol, EuroJournals Publishing, Inc. 2009, ISSN 1450-216X Vol.32 No.4 (2009), pp.444-454 http://www.eurojournals.com/ejsr.htm

[8]   Zainab R. Zaidi, Brian L. Mark: "A Distributed Mobility Tracking Scheme for Ad-Hoc Networks Based on an Autoregressive Model". The 6th International Workshop of Distributed Computing, Kolkata, India (2004) 447(458)

[9]   Abdullah, SohailJabbar, ShafAlam and Abid Ali Minhas, "Location Prediction for Improvement of Communication Protocols in Wireless Communications: Considerations and Future Directions", Proceedings of the World Congress on Engineering and Computer Science 2011 Vol. II WCECS 2011, October 19-21, 2011, San Francisco, USA

[10] Gunnar Karisson et al., "A Mobility Model for Pedestrian Content Distribution", SIMUTools '09 workshops, March 2-6, 2009, Rome Italy.

[11] John Tengviel, K. A. Dotche and K. Diawuo, "The Impact of Mobile Nodes Arrival Patterns In Manets Using Poisson Models", International Journal of Managing Information Technology (IJMIT), Vol. 4, No. 3, August 2012, pp. 55 – 71.

[12] Martin J. Fischer and Carl M. Harris, "A Method for Analysing Congestion In Pareto and Related Queues", pp. 15 – 18.

[13] K. Krishnamoorthy, "Handbook of Statistical Distributions with Applications", University of Louisiana at Lafayette, U.S.A. pp. 257 – 261.

[14] Kyunghan Lee, Seongik Hong, Seong Joon Kim, Injong Rhee and Song Chong, "SLAW: Self-Similar Least-Action Human Walk", published in the Proceedings of the IEEE Conference on Computer Communications (INFOCOM), Rio de Janeiro, Brazil, April 19–25, 2009.. PP. 855 – 863.

11

COORDINATING DISASTER RELIEF OPERATIONS USING SMART PHONE / PDA BASED PEER-TO-PEER COMMUNICATION

Debanjan Das Deb[1], Sagar Bose[2] and Somprakash Bandyopadhyay[3]

[1]Department of Computer Science & Engineering,
B. P. Poddar Institute of Management & Technology, India
`debanjandasdeb@gmail.com`
[2]PervCom Consulting Pvt. Ltd., India
`sagar@pervcomconsulting.com`
[3]Indian Institute of Management Calcutta, India
`somprakash@iimcal.ac.in`

ABSTRACT

During any post-disaster period, the availability of the Internet is ruled out in most cases, mobile phones are only partially usable in some selected regions. Candidate devices for maintaining minimal services are mostly expensive satellite phones or specialized point-to-point radio communication systems. As communication systems become crippled, so do the management of the relief operations. One of the common problems during disasters is that the rescue and relief operations are not well-coordinated. For this reason, there is a need for a system that will help in the efficient distribution of rescue and relief to disaster-affected areas. The objective of this paper is to propose a smart-phone/ PDA based disaster management system based on peer to peer communication only and supporting disconnected operation.

KEYWORDS

Peer-to-Peer Communication, Opportunistic Network, Delay-tolerant network

1. INTRODUCTION

Any large-scale disasters like flood and cyclone have severe impact on communication infrastructure. Services that are relied on for everyday communications (e.g., cell phone / internet connectivity) immediately become non-functional in emergency situations due to the failure of the supporting infrastructure through both system damage and system overuse [1]. In contrast to the vulnerable fixed network infrastructure, it is very likely that battery-powered wireless personal mobile communication devices (PDA, cell-phones) will survive in disasters. Additionally, even more mobile devices will be brought to the scene by relief workers. Currently, those devices are having powerful processors and high storage capacity with GPS and multi radio interfaces (Cellular, Wi-Fi, Blue-tooth). Such devices are, therefore, promising candidates to contribute in forming peer-to-peer wireless network structure to support disaster communication.

However, end-to-end connectivity can never be assumed in this kind of scenario and long disconnections are the rule. Thus, in this context, the devices (PDA, cell-phones) spread across the environment by relief workers form peer-to-peer network, albeit disconnected. In this type of networks, the mobility of devices is an opportunity for communication rather than a challenge. Mobile nodes communicate with each other in peer-to-peer mode.

The objective of this paper is to develop a dynamic, virtual star topology with static central control station as root node and static shelter points as end-nodes (figure 1). The connectivity between root node and each of the end-nodes is achieved using mobile volunteers opportunistically as message ferry. At the same time, the mobile volunteers also exchange information among one another in a peer to peer mode, thus integrating the field information intelligently and autonomously using auto-configurable mobile-phone-based peer-to-peer communication. Basically mobile volunteers act like end devices. The major focus of this study is to come up with a framework for post-disaster resource requirement analysis, resource allocation & distribution, which involves on-going determination of what resources are needed, what resources are present, what resources need to be acquired and how long will it take for them to arrive. Using peer-to-peer communication, eventually the entire network will appear as virtually connected, delay-tolerant network with every node knows the approximate information about other nodes, including resource need. Thus, the system enables coordinated relief operation, where no shelter will be having excess resource, nor it will be under-resourced.

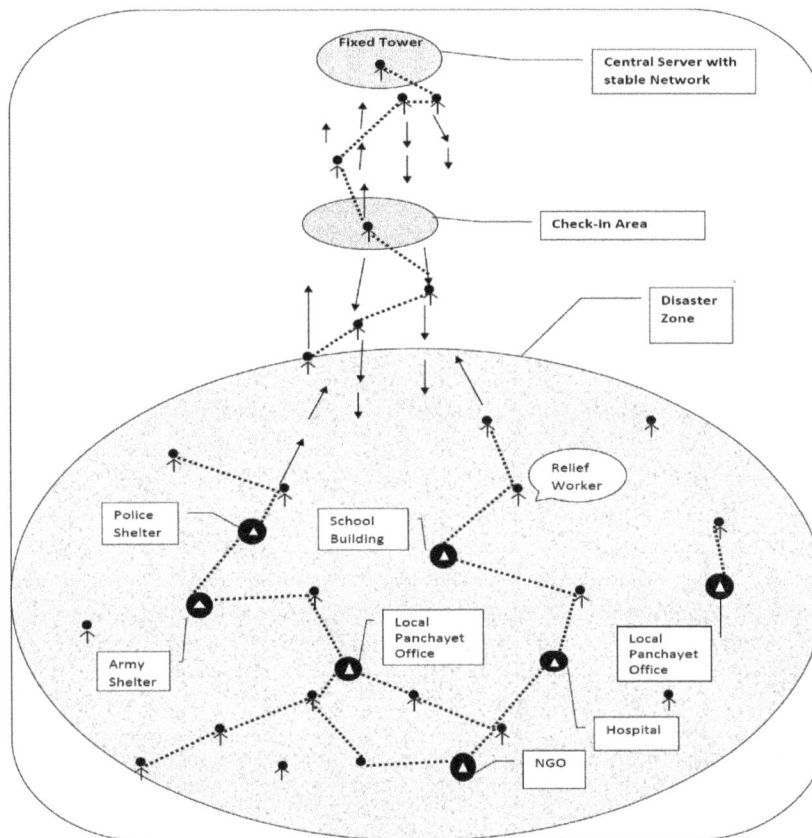

Figure 1. Framework for post-disaster communication between Central Server, Shelter and Volunteers

2. RELATED WORK

In disasters, it is necessary to have established procedures for obtaining additional resources when they are needed. However, indiscriminate requests for resources can be detrimental. Many disasters are complicated by the over-response of resources, and this can greatly complicate the already difficult problems of coordination and communication. Procedures for pinpointing the specific types and numbers of resources needed are helpful in making the disaster response more manageable. Disasters with multiple impact sites and large, complex disasters often call for an emergency operations center (EOC) (Erik, 1989) [4]. The EOC or Central Control Station is usually established away from the disaster scene, often near governmental offices in contrast to the command post, which is concerned with activities at the scene, the EOC establishes priorities for the distribution of resources among the various sites, and handles off-incident concerns. This is basically a centralized system for controlling disaster relief management. However, a major concern is the connectivity between EOC and the shelter points in the field.

"VSAT technology", which is particularly useful when terrestrial infrastructure has been destroyed, provides a powerful tool to mitigate damage incurred by disasters (Gorp, 2008) [5]. The deployment of VSAT is typically seen as the last option for obtaining access to larger communications network and the Internet. It is often necessary for organizations operating in remote areas. However, VSAT Technology is expensive and its deployment is complex in nature. Technical support is critical for VSAT deployment after disasters, because in the disaster area the people are much more focused on helping people than messing with technology. VSAT is often deployed in the context of establishing new field offices; and thus is primarily deployed for development purposes, rather than for emergency response, or direct post-disaster relief.

Multiple-Criteria Decision Technique, together with a coordination strategy and a communication strategy, have been deployed in order to assure that the decision making process has the appropriate information upon which to perform the resource distribution (Silvia, 2003) [6]. The coordination strategy allows distributing resources to the victims that need the most urgent rescue. The communication strategy emphasizes information flow concerning disaster victims. The role of the moving agents is to gather information about victims and the role of the fixed agents is to pass on this information to the fixed stations. Here, some communication infrastructure has been assumed. However, if there is no communication at all between agents, then results show that moving agent get lost in the rescue scenario and cannot find victims.

Cluster-based Communications System focuses on the vital need for providing communications facility to the victims, immediately after the disaster and prior to the arrival of rescue teams (Sonia, 2009) [7]. The proposed novel approach in emergency communications enables survivors to communicate among themselves and help each other. Here, the idea of a self organizing ad hoc network is put forward, which makes use of available network resources formally occupied by the destroyed and / or damaged telecommunication infrastructure. In post-disaster scenario, mobile nodes establish a self-organizing / ad hoc network which provides critical level of communications among disaster victims needed at that time and consequently tries to merge with some surviving telecommunications infrastructure and / or network deployed by rescue teams. But, depending on the nature of disaster, a connected ad hoc network may never be formed, and thus communication among victims cannot be established.

Prototypical Crisis Information Management System (Iannella, 2007) [8] is a conceptual framework to support two challenges: incident notification and resource messaging. The mantra for Crisis Information Management Systems is to "deliver the right information to the right people in the right format in the right place at the right time". These five variables, coupled with the stress of a major disaster, make coordinated information management one of the greatest challenges for the disaster coordination sector. However, this system assumes that a strong ICT infrastructure is in place in order to enable cooperation between a large number of organizations and integration with these organizations. A similar collaborative mechanism has been proposed by Marrella (2011) [9], where the communication is executed on top of mobile networks. However, the collaboration strictly depends on the possibility that operators and their devices can communicate with each other using some communication backbone.

Starting from collected user requirements and their generalization, WORKPAD architecture (de Leoni, 2007) [10] is based on a 2-levels peer-to-peer (P2P) paradigm: the first P2P level is for the front-end and the latter level is for the back-end. The need of such two P2P levels arises from the analysis of user requirements, as there exist back-end central halls where the chiefs of involved organizations are located, as well as several front-end teams which are sent to the affected area. This architecture is based on a stable centrally connected network where all the information percolates and centrally managed. If the entire connection disrupted then the efficiency level will decrease drastically.

3. SYSTEM ARCHITECTURE

3.1. System Description

During disaster, victims of a village / locality normally take shelters in groups in some nearby safe areas (for example, some school buildings, Army-Police camp and Panchayet Offices in some highland areas, etc.). Therefore, coordination of post-disaster relief operations and distribution of resources are shelter-centric (figure 2) through the relief-workers working at those shelters (figure 3). These relief workers are exchanging their presence, field information and shelters requirements by using their smart phones in an opportunistic network. These shelters are operated by a Shelter Coordinator who will periodically coordinate with nearest relief workers autonomously for analysing field need and informing shelter need through smart phone. Another role of shelter coordinator is to control the navigation pattern of nearby relief workers with an objective to spread the information to other shelters and eventually to Central server. Thus the information percolates from lower layer to upper layer through relief workers.

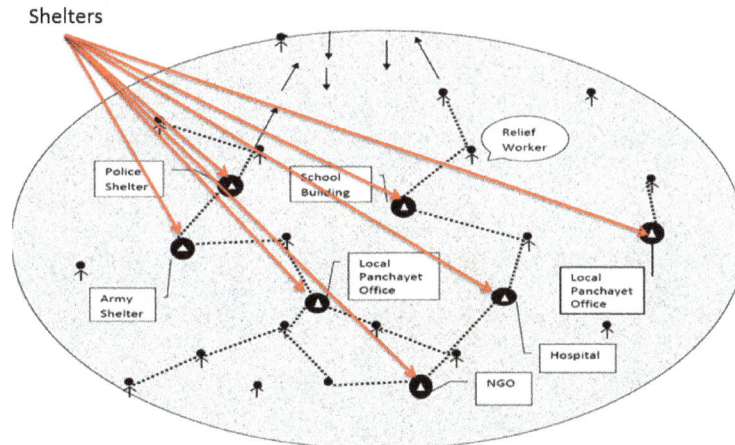

Figure 2. Shelters in the affected zone

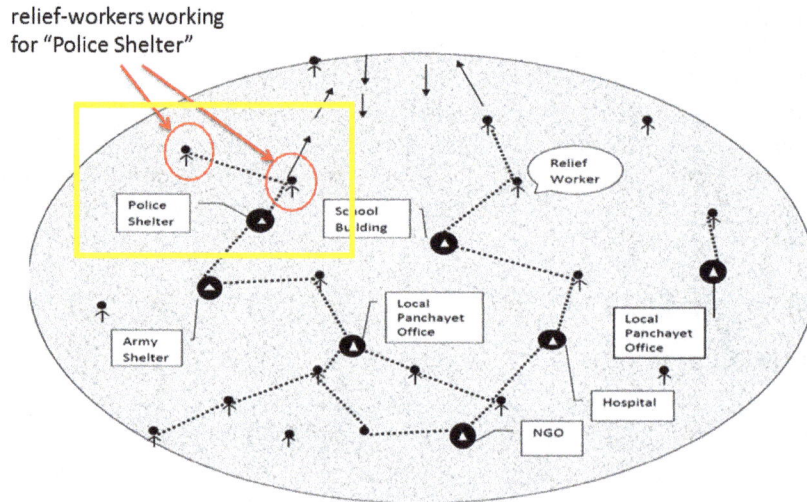

Figure 3. Relief workers working at shelters

The shelter coordinator redirects relief workers by evaluating the field requirement with a primary intention to send at least a few workers towards central server. Once relief workers are get connected with another relief worker, the information exchanges and the modified information percolated in multi hop. Central server also sends few relief workers with relief materials towards disaster area through warehouse. This central server is established in an area where it gets proper communication network. It receives information from relief workers and sends its directions to the relief worker who is nearer to central server and then that relief worker exchange that information / resource (relief materials) to its nearer worker and by this way the total information spreads out all over the affected area (figure 4).

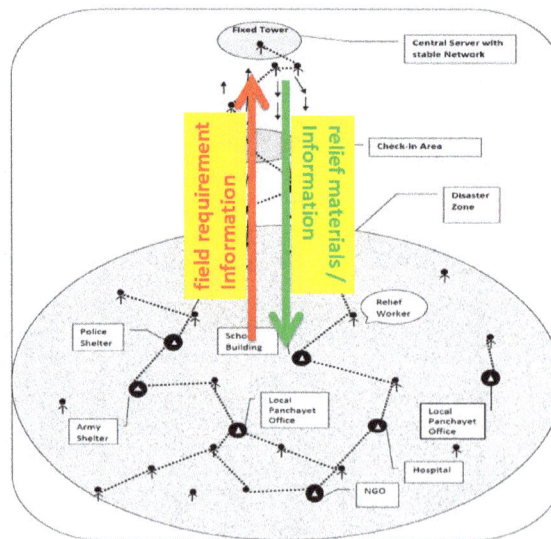

Figure 4. System coordination by Central Station

3.2. Formation of Smart phone based Opportunistic Network

In the decentralized Disaster Management scenario, relief workers are exchanging their presence, field information and shelters requirements by using their smart phones in an opportunistic network. Opportunistic network as one type of challenged networks where network contacts are intermittent or where link performance is highly variable or extreme. In such a network, there does not exist a complete path from source to destination for most of the time. In addition, the path can be highly unstable and may change (make or break) quickly. Therefore, in order to make communication possible in an opportunistic network, the intermediate nodes may take custody of data during the blackout and forward it when the connectivity resumes. Thus, in this context, Opportunistic Network framework [2, 3] provides potential platform for information communication. In opportunistic networks, the devices (PDA, cell-phones) spread across the environment by relief workers and form the network. In this type of networks, the mobility of devices is an opportunity for communication rather than a challenge. Mobile nodes communicate with each other even if an end-to-end route connecting them never exists. Any node can opportunistically be used as the next hop, if it is likely to bring the message closer to the destination(s).

Currently, those devices or Smart Phones are having powerful processors and high storage capacity with GPS and multi radio interfaces (Cellular, Wi-Fi, Blue-tooth). These Smart Phones are not only useful for making telephone calls, but also adds features that might be found in a personal digital assistant or a computer. Smartphone also offers the ability to send and receive messages on peer to peer basis using GPS or Wi-Fi, Blue-tooth technologies and edit Office documents. Such devices are, therefore, promising candidates to contribute in forming ad-hoc wireless network structure to support disaster communication.

In a seminal paper in Physical Review Letters, Vicsek et al.[13, 14] propose a simple model of n autonomous agents moving in the plane with the same speed but with different headings. Each agent's heading is updated using a local rule based on the average of its own heading plus the headings of its "neighbors." In their paper, Vicsek et al. demonstrated that the nearest neighbor rule can cause all agents to eventually move in the same direction despite the absence of centralized coordination and despite the fact that each agent's set of nearest neighbors change with time as the system evolves. Other studies also indicate that multi-agent systems that interact through nearest-neighbor rules can synchronize their states regardless of the size of communication delays [15]. We have applied this concept in our system and the performance evaluation results indicate the effectiveness of our approach.

3.3. Decentralized Information Assimilation & Coordination

In this scenario the information generated on peer to peer basis between relief workers and static shelter points by using opportunistic network. This information percolates from lower layer to upper layer through shelter belonging to same cluster (figure 5).

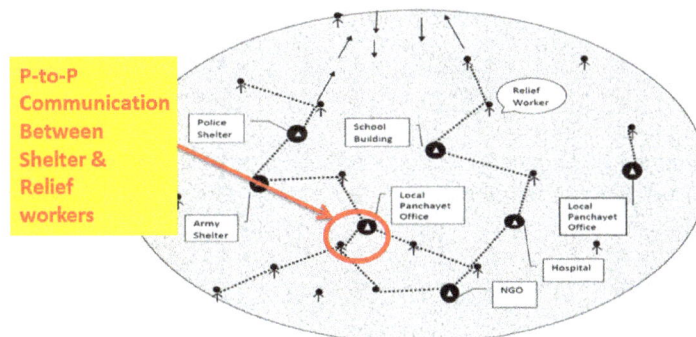

Figure 5. Peer-to-peer communication (all dotted lines) between shelter and relief workers

The Shelter leaders redirect relief workers by evaluating the field requirement with a primary intention to send at least a few workers towards central server. Once relief workers are get connected with another relief worker, the information exchanges and the modified information percolated in multi hop (figure 6).

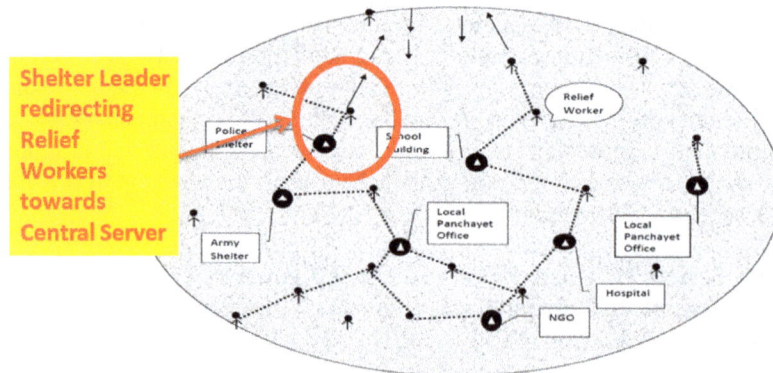

Figure 6. Shelter Leader / coordinator redirecting relief workers towards Central Server

Central server also sends few relief workers with relief materials towards disaster area through check-in-area. This central server is established in an area where it gets proper communication network. It receives information from relief workers and sends its directions to the relief worker who is nearer to central server and then that relief worker exchange that info to his nearer worker and by this way the total information spread out to entire area (figure 7).

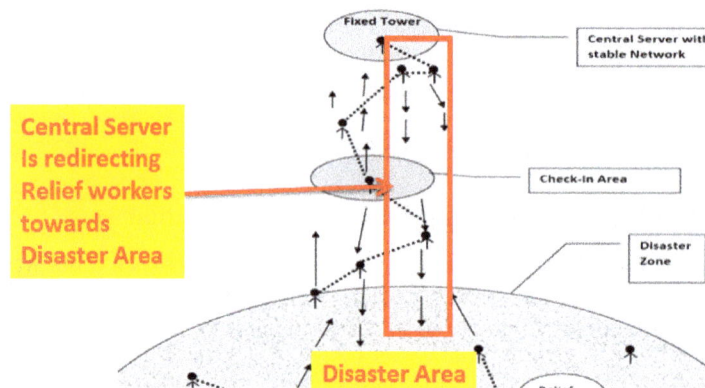

Figure 7. Central Server is redirecting relief workers towards disaster area

The static shelter coordinator sends information by using Smart Phone to nearer relief worker. This information contains sender ID, location and sending time which will help to review the information time and identification status. Relief worker also has the option to forward that received information from shelter coordinator to another relief worker by updating its part of information like:

- Id

- Location of work

- Status of work (whether pending or will take time)

- Navigation status (if the shelter coordinator has instructed to navigate to central server or any other shelter)

- Infrastructure information

- Information related to difficulties faced

- Necessary resource requirement

Shelter coordinator also spread out information of:

- Infrastructure towards his shelter with location specification.

- Victim's status like

  o   how many of them are still alive,

  o   how many are required hospitalization

- Details of resource requirement and present resource status of the shelter.

- Navigation request: At first it will judge shelter's resource requirement (not only its shelter but also other neighbouring shelters) then depending on the situation it will redirect relief workers towards shelters or towards central server.

Central server also sends its instruction or information in multi hop through relief workers who are approaching towards disaster zone by spreading information through smart phone on Peer to Peer (P2P) basis. It also specifies the sending time, ID and location in its message. Relief workers are not able to modify that information. They can only exchange information by forwarding that message coming from Central server. In this case, relief workers have the option to modify the forwarding time, location and its Id only. Depending on that, any receiver of that message can easily filter the information coming from central server on time and location basis. The main contents of message from central server are

- Resource sending status like quantity of resources have already sent for specific shelter

- Request towards any other shelter to navigate its resources towards that specific shelter

- Request for emergency services like the necessary initiatives should be taken by Police, Army, Doctors, NGO professionals, and Fire Bridget for any specific location

- Instruction for shelter coordinator

This is a random process by which the total information percolates to entire disaster region. It will create a standard data base on which a specific software model can be established for retrieving required information by filtering this. Relief workers will get an option to save the received messages on its phone and by saving all these information, there will create a broad data base for them where the information will automatically be updated by saving it and repeated information will automatically get erased.

Central server will get the entire disaster zone related information from different relief workers periodically and will create a broad data base depending on that information in its system. This data base will be circulated to outside world for their knowledge and for asking necessary support from outsiders. The formation of check in area is an activity of central server which will work as a central wire house for keeping relief resources systematically and send them periodically towards disaster zone when these are required. This check in area will work as a manager for the distribution of resources properly. It is an area where initially all relief workers will accumulate and then will spread out to whole disaster zone. To reduce the pressure on Central Server this Check-in-Area will work as a coordinator between central server and main disaster zone. The segregation of work will standardize the pattern of distribution and communication in a decentralized disaster environment. However, there is no fixed network established in the disaster area.

4. SIMULATION

4.1. Simulation set up

We are interested in studying a post disaster scenario in which the resources and aid can effectively be distributed to the shelters being formed amidst the disaster affected area. The environment consists of few entities of our concern that are essential in studying the scenario. They are the volunteers or relief workers which co-ordinately achieve some tasks and form the only medium for dispersion of information throughout the affected area. There are fixed locations or relief camps at certain points known as shelters where victims take shelter and it is their needs that the volunteers ensure to fulfil. There is a fixed location outside the affected site known as the Check-in-Area (CA) from which coordinated and centralized relief operations are being monitored. There is no direct network connectivity among CA and shelter points. Next we have another static entity known as the fixed Central Server which acts as a resource pool and a medium to connect to the outside world. CA and Central Server also have no network connectivity.

Since there are few categories among them which performs the same tasks and have the same attributes, the best way to represent them in a similar environment is to incorporate them as agents and a very powerful toolkit that comes handy in this perspective is REPAST Symphony. Recursive Porous Agent Simulation Toolkit (REPAST) is a free and open source agent-based modelling toolkit that offers users with a rich variety of features [11, 12, 13, 14]. This simulated environment is written in JAVA.

4.2. Simulation parameters

- The proposed scheme is evaluated on a simulated environment by varying key parameters related to the model to study and estimate the efficiency and uniformity of resource distribution against time-ticks.

- In the simulation, the environment to be studied is assumed to be an area of 10 k.m. x 10 k.m. in which mobile volunteers are initially distributed randomly. However, the affected site is considered to be a rectangular area of size 10 k.m. x 4 k.m. in which all the shelters are present.

- For all cases we have kept the shelter count constant and it equals 10. All the simulated actions are associated with time-ticks.

- In the Repast platform, "ticks" are considered as a unit of time; we have assumed 1 tick equals 1 minute and hence the speed of movement of individual volunteers is considered to be 100 meters per minute.

- In this environment, relief resources like food, medicine, clothes, etc. is considered as an aggregated integer quantity, called resource-count.

  o The positive resource-count indicates that resources are present in excess.

  o The negative resource-count indicates that there is a demand of resources at the shelter and no resource is present.

  o If resource-count equals zero, it means that neither there is a demand nor there is any surplus of resource at that shelter.

4.3. Simulation environment overview

- The volunteers move randomly at the beginning.

- As indicated, each shelter has either positive or negative resource. Each shelter autonomously decrements the resource-count periodically (self-decrement of resource-count), indicating that victims present at the shelters are consuming resources and the net resource-count at each shelter decreases randomly with time.

- When a volunteer visits a shelter whose resource-count is negative (i.e. when there is a demand), then it goes directly to the CS, to give that information and to get the needed resource physically. In this way, the CS gets the information of resource-need and resource distributed to all the shelters. The shelter ensures that no other volunteer visits with the same piece of information to the CS.

- When a volunteer goes to the CS with negative resource information, then at separate times of the simulation, two cases might arise :

  I.   At the initial phase of the simulation, the CS may start giving out resources from itself in a little excess quantity than required until there are enough shelters being discovered with excess resources.

  II. After there are plentiful of references to the shelters being discovered, only then the CS assigns tasks to the respective volunteers involved to transfer resources from a nearby shelter having positive resources to the shelter with negative resources.

- It is assumed that each volunteer is carrying PDA / smartphone with Wi-Fi / Bluetooth interface for peer-to-peer communication. It is also assumed that the cellular communication infrastructure is not functioning and volunteers are exchanging information only in peer-to-peer mode.

- At first volunteers have no knowledge about other volunteers and location of shelters. They move randomly in the disaster area to discover shelters and other volunteers. When a volunteer roaming randomly comes near a shelter or the shelter is within the vision of the volunteer, then the volunteer checks its own queue (the queue storing shelter IDs).

  o If that particular shelter is present in the queue, it signifies that this volunteer had recently visited that shelter. So, no need to again visit that shelter until recently but exchange information with that shelter. And again the volunteer starts roaming randomly.

o   If there is no occurrence of the shelter in its list then the volunteer approaches the shelter and waits there until the tag of the shelter (previously described) becomes false.

- The same procedure is performed while another roaming volunteer comes under the vision of this volunteer. Both exchange information. In this case, if the occurrence of meeting with the other volunteer is present in this volunteer (queue) then, it will not exchange information because, it is probable that more or less the information might be redundant.

    o   Before going further it is needed to describe the data structures and the type of data that a volunteer, shelter and Central Station keeps with them. They are discussed below :

    I)   Data kept by a volunteer:

    - Name (or ID) and location of the shelters known to it.

    - Name (or ID) of other volunteers present and known only to this volunteer.

    II)   Data kept by a shelter:

    - Name (or ID) and location of other shelters present in the affected area known to it.

    - Name (or ID) of other volunteers present and known only to this shelter.

    - It keeps all the timestamps history of when (at what time-tick) a specific volunteer had visited this shelter (required to track a volunteer).

    III)   Data kept by the central station:

    - Similar to shelter and volunteers it also keeps a list of shelters and volunteers.

    - Name (or ID) and location of all shelters present in the affected area known to it. Initially, it is empty and is updated by the volunteers.

    - Name (or ID) volunteers present in the affected area. Initially, it is empty and is updated by the volunteers.

    - A shelter's resource status list that stores the status of resources at the shelters known to the CS through interaction with the volunteers.

- While roaming, the volunteer also checks the status of the shelter (a Boolean flag); two cases might occur at this moment:

    o   If the flag is found true (it means some other volunteer is working for this shelter to fetch and transfer resources), then it waits for a certain number of time-ticks describing the fact that it is both waiting for the flag to get false and it is performing some work at the shelter. If the total time for waiting at the shelter elapses then it moves on to the next shelter in its list i.e. it moves to the next unvisited shelter whose reference (name & location) is present with it. It also keeps a check on this shelter's entry in its own table, so that it can pay a visit once again in the near future. But before leaving the shelter, it ensures to exchange any new information (present with it) with the shelter.

- o If the flag is found false, then no one is working for this shelter and hence it gets an opportunity to work for this shelter. It then immediately makes the flag true and checks the status of the resource present at this shelter. If it is greater than equal to 0 then no need of going to the CS because there is no demand of resources. Instead leave this shelter and go to the next shelter in its list.

- Again if the resources are found negative, then it immediately goes to the CS to fetch resource for it, whereby the CS gives it a task to perform.

- A volunteer does not keep the record of the resources of previous entries of a shelter. It only keeps the status of the resources of the shelter that it is involved presently. A volunteer at a time associates with only one shelter. When a volunteer is involved with a shelter, at that moment only that volunteer can perform the tasks related to the shelter. No other volunteer can interfere to take charge of the tasks; but surely it can exchange information. A shelter similar to a volunteer is also associated to a single volunteer for resource updating at a time.

- When a volunteer comes to deliver negative resource information from a shelter to the CS, the CS first updates any new piece of information from the volunteer and then takes the recent resource status (negative) of the shelter it is coming from.

- After updating its own list of resource status of the shelters, it then runs an algorithm which finds out a nearby shelter (to the concerned shelter) having excess resource. If there is no such shelter with excess resource within the vicinity of the concerned shelter, then it gives out resources from itself and hence directs the volunteer with resources directly to the shelter in demand.

4.4. Assumptions for simulation

- There is a common synchronised watch available to all volunteers.

- The consumption of resources is instantaneous i.e. the resources are consumed immediately when they are being transferred to a shelter.

- The shelters are assumed to be formed before the central station can take notice of them and they have to be discovered by the volunteers.

- Volunteers carrying PDA / Smart-phone, supporting query / exchange information autonomously

4.5. Simulation results

- Figure 8 shows the individual resource status of the shelters in the affected area w.r.t. time-ticks.

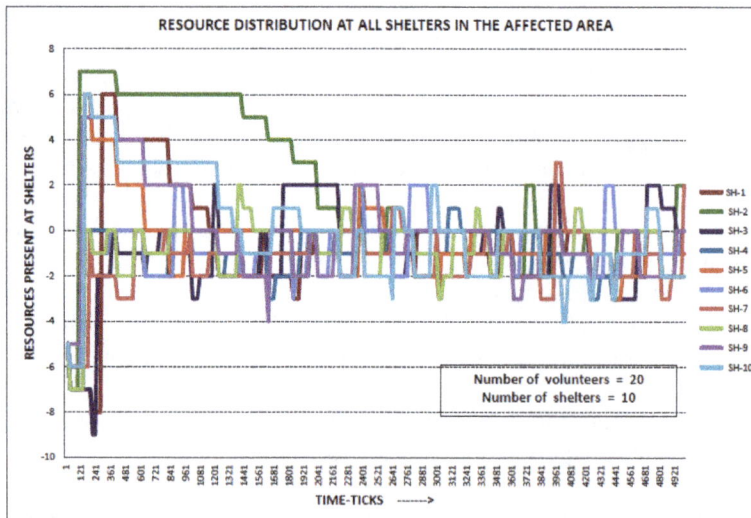

Figure 8. Resource distribution at all shelters in the affected area

- This graph is a direct consequence of the simulation with number of volunteers and shelters set at 20 and 10 respectively.

- The shelters possess self-decrement of resources which means there is a demand for resources at the shelters. The central station is assumed to be situated at an approximate distance of 3 kilometres from the affected area.

- It is deduced from the graph that initially there is a huge demand and it increases with the time-ticks till 250 to 300 ticks. Before this time the volunteers were busy discovering the unknown shelters that were formed previously in the affected area.

- The transferring of resources to a shelter is reflected in the graph when the net resource at a shelter increases from negative to 0 or to the double of the demand.

- The most interesting part is that as time increases, the demands of the shelters are replenished quickly as and when noticed by the volunteers. The graph also shows that majority of the shelters gets their demands fulfilled before their resource status reaches -3. Most are replenished quickly at -1 and -2 before their demand increases further.

- From this graph we conclude that with the increase of demand for resources at shelters, the volunteers deliver the information to the central station, which in turn looks after the fact that no shelter starves for resources for a long time (apart from the time needed to fetch and transfer the resources). This is seen in the later phase of the graph where the initial fluctuating pattern is brought down to much more linear pattern, where the resources at any shelter is neither too large (in excess) nor is devoid of it for a long period of time.

- Figure 9 represents a screenshot of one of the simulation in REPAST taken at time tick = 250. The number of volunteers working in the field and the number of shelters present are set to 20 and 10 respectively. The central station is situated at an approximate distance of 3 kilometres from the affected area.

- In figure 9 we see that the volunteers (represented by small blue dots) moving towards and away from the Check-In-Area. Some volunteers had changed their color to red, which signifies that they are moving out from the Check-In-Area carrying resources for their respective shelters. The agents which oscillates between the Check-In-Area and the Fixed Tower (represented by small black dots), belongs to the Check-In-Area and are seen to move towards the Fixed Tower. They are transferring any new static information present at the Check-In-Area to the Fixed Tower.

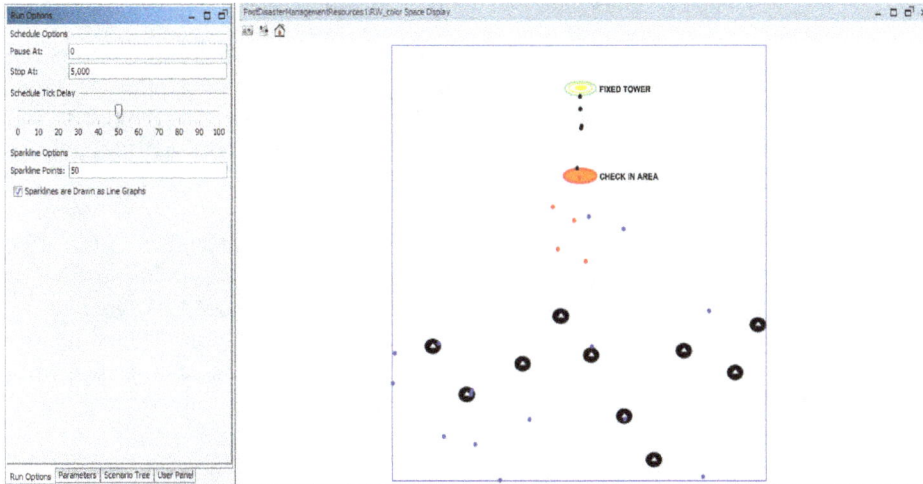

Figure 9.   Screenshot of a simulation taken at time tick = 250. The volunteer and shelter counts at the field is set to 20 and 10 respectively. The boundary specifies our area of interest and measures 10 kilometres x 10 kilometres.

- Next we have figure 10 which shows the perception of resources by the central station at the affected area after it supplied the resources vs. the actual resources present at the area.

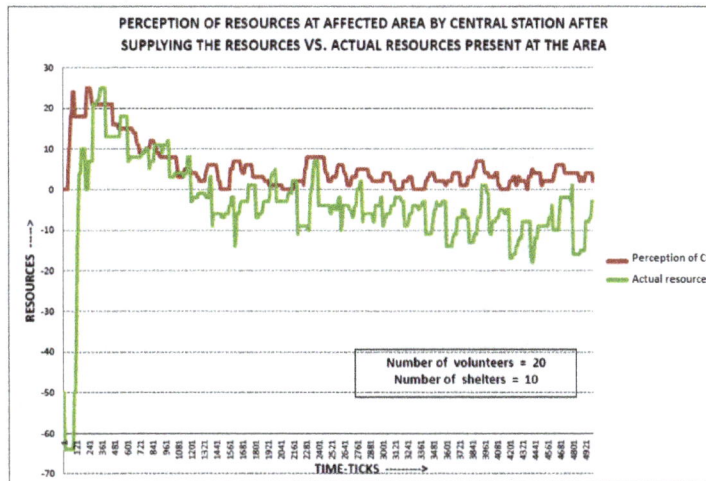

Figure 10. Perception of resources at affected area by central station after supplying the resources vs. Actual resources present at the area

- The number of volunteers working in the affected area and the number of shelters are 20 and 10 respectively. The distance between the central station and the affected area is moderate. This graph refers to the same simulation with the same conditions as graph of figure 8.

- Initially the central station assumes that the sum total of all resources at shelters (if any) is 0, before they are discovered. As the volunteers discover the shelters and notify the central station of the demands, the Central Station updates its own list accordingly.

- It is to be remembered that this graph is a result of post supplying the resources by the central station. From this graph (the red line) it is seen that the perception of resources is always positive and hence understood that the central station tries its best to supply and distribute resources throughout the affected area in an effective way by applying its algorithm on its list.

- However, the other graph (the green line) showing the actual resources in the affected area depicts a deviation from the central station's viewpoint. But this graph almost resembles the perception graph. This is due to the delay introduced in overcoming the distance of separation between the shelters and the central station. This similarity of pattern between both the graphs concludes that even if the central station is placed at a distance of 3 kilometres from the affected area, the plan is working out as planned by the central station.

- Figure 11 represents the variations of the sum total of all resources present in the affected area w.r.t. time-ticks by varying the volunteer count.

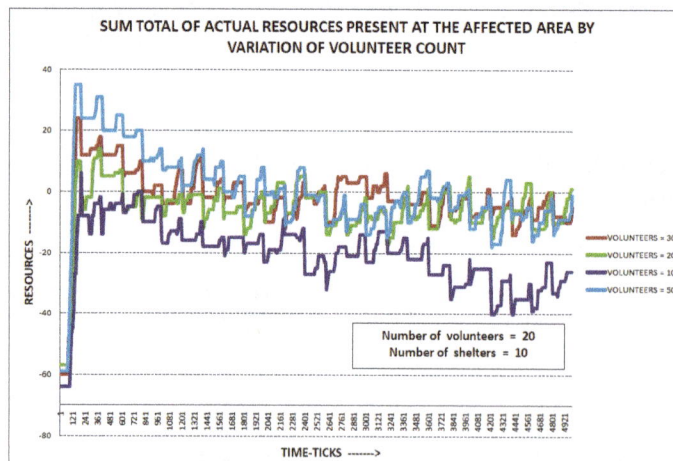

Figure 11. Sum total of actual resources present at the affected area by variation of volunteer count

- The graphs are plotted with volunteer counts of 10, 20, 30 and 50 respectively. The shelter count remains constant for all the simulation and equals 10.

- The simulations include the self-decrement of resources at shelters and the central station being placed at a distance of 3 kilometres from the affected area.

- From the graph with volunteer count = 10, we see that due to less number of volunteers, the volunteers do not get the opportunity of visiting the other shelters more frequently and are either busy searching for new shelters or is busy in a task. With less number of volunteers the shelters get starved.

- On the other hand we notice that the total resource distribution of the graphs with volunteers =20, 30 and 50 falls within the same range i.e. the volunteer count values yields almost the same result in the long run.

- This justifies the fact that at some point, even if the volunteer's counts are increased, the net resource distribution at the affected area remains same. As depicted from the graphs, the optimum volunteer count in the affected area can be assumed to be 20. Even if volunteers are increased beyond this value, the net distribution of resources will not be affected much.

- Next we have figure 12 which shows the convergence of static information (i.e. the number of volunteers and the number of shelters present) throughout the affected area w.r.t. time-ticks. These graphs are plotted to study the variation of the percolation of this information through every shelter and volunteer in the area.

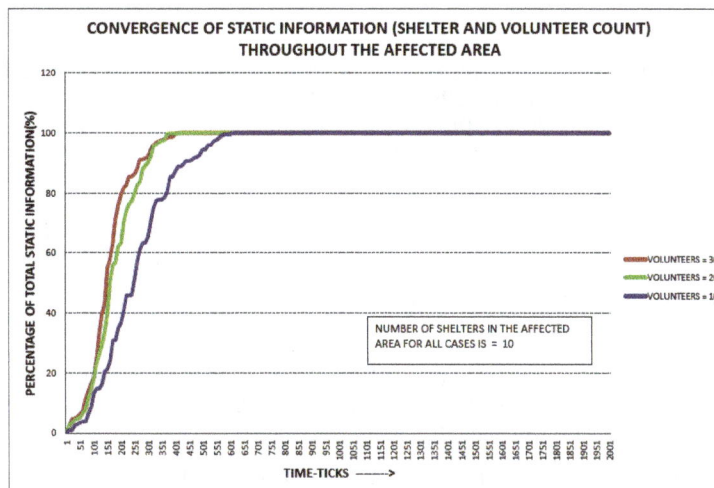

Figure 12. Convergence of static information (shelter and volunteer count) throughout the affected area

- The number of shelters is kept constant at 10 and the volunteer counts are varied by values 10, 20 and 30. The Y-axis represents the percentage of convergence of the total static information. For this we have taken the percentage of the information every volunteer and shelter carries individually and then take an average to find the aggregate percentage. The resulting graphs are inverted exponential graphs and the study reveals that as the volunteer counts increases, the time required to percolate the information throughout decreases to a certain point beyond which increasing the volunteer count yields the same result, i.e. the 100% convergence point coincides for values beyond 20.

5. CONCLUSIONS

In this paper, we have proposed an effective scheme of providing a coordinated post disaster relief operation. Assuming the absence of a network infrastructure, we tried to incorporate the notion of opportunistic network by using peer-to-peer communication between relief workers. We investigated the schema by varying key parameters that affects the performance of the system. The simulation results of the system shows that even if the entire coordinated operations are centralized and is monitored far away from the affected area, the operations yields positive results in the overall resource present at the affected area. We have also shown that the perception of resource pattern bears a similar trend to the actual resource pattern at the affected area proving the fact that information is percolated and perceived properly. Furthermore, we have found out the optimal relief worker count and proved that this optimal value has a direct impact on the convergence of static information throughout the affected area as information is dispersed quickly. We have also shown that resources are effectively distributed as and when demand arises at the shelters. The findings of this paper suggests that even in the absence of a network infrastructure, using peer-to-peer communication and message passing can eventually form an intermittent opportunistic network which performs effectively even if the key parameters are varied widely. This study thus reveals the applicability of the scheme in PDAs and smart-phones which can be used to form peer-to-peer wireless network during relief operations.

REFERENCES

[1]   H. Luo, R. Kravets, T. Abdelzaher, The-Day-After Networks: A First-Response Edge-Network Architecture for Disaster Relief, NSF NeTS FIND Initiative, 2006-2010. <http://www.nets-find.net/Funded/DayAfterNet.php>.

[2]   Marco Conti, Mohan Kumar: Opportunities in Opportunistic Computing. IEEE Computer 43(1): 42-50 (2010)

[3]   Conti, M.; Giordano, S.; May, M.; Passarella, A.; From opportunistic networks to opportunistic computing, IEEE Communications Magazine, IEEE , Volume: 48 Issue:9, Sept. 2010

[4]   [Erik Auf der Heide, 1989]. Disaster Response - Principles of Preparation and Coordination, Publisher: Elsevier-Medical, 1989, p. 303

[5]   [Gorp, 2008]. Annemijn van Gorp et al, "VSAT Deployment for Post-Disaster Relief and Development: Opportunities and Constraints for Inter-Organizational Coordination among International NGOs", 17th Biennial Conference, 2008.

[6]   [Silvia, 2003], Suárez Silvia, López Beatriz, De la Rosa Josep Lluis. "MCD Method for resource distribution in a large-scale disaster". In: "X Conferencia de la Asociación Española para la Inteligencia Artificial" CAEPIA 2003, volumen II, pág. 261-264, San Sebastián, Spain, November 11-14 of 2003.

[7]   [Sonia, 2009]. Sonia Majid, Kazi Ahmed, "Cluster-based Communications System for Immediate Post-disaster Scenario", Journal of Communications, Vol. 4, No. 5, June 2009.

[8]   [Iannella, 2007]. R. Iannella and K. Henricksen, "Managing Information in the Disaster Coordination Centre: lessons and opportunities", in ISCRAM 2007 Conference, Delft, The Netherlands, May 2007.

[9]   [Marrella, 2011]. Andrea Marrella et al, "Collaboration On-the-field: Suggestions and Beyond", in ISCRAM 2011 Conference, Lisbon, Portugal, May 2011.

[10]   [de Leoni, 2007]. de Leoni et al., "Emergency Management: from User Requirements to a Flexible P2P Architecture", In Proceedings of ISCRAM 2007, 2007.

[11]   Howe, T.R., N.T. Collier, M.J. North, M.T. Parker, and J.R. Vos, "Containing Agents: Contexts, Projections, and Agents," Proceedings of the Agent 2006 Conference on Social Agents: Results and Prospects, Argonne National Laboratory, Argonne, IL USA (September 2006).

[12]   North, M.J., P. Sydelko, J.R. Vos, T.R. Howe, and N.T. Collier, "Legacy Model Integration with Repast Simphony," Proceedings of the Agent 2006 Conference on Social Agents: Results and Prospects, Argonne National Laboratory, Argonne, IL USA (September 2006).

[13]   Parker, M.T., T.R. Howe, M.J. North, N.T. Collier, and J.R. Vos, "Agent-Based Meta-Models," Proceedings of the Agent 2006 Conference on Social Agents: Results and Prospects, Argonne National Laboratory, Argonne, IL USA (September 2006).

[14]   Tatara, E., M.J. North, T.R. Howe, N.T. Collier, and J.R. Vos, "An Introduction to Repast Modelling by using a Simple Predator-Prey Example," Proceedings of the Agent 2006 Conference on Social Agents: Results and Prospects, Argonne National Laboratory, Argonne, IL USA (September 2006).

[15]   A Survey of Opportunistic Networks, Chung-Ming Huang, Kun-chan Lan22nd International Conference on Advanced Information Networking and Applications – Workshops, 978-0-7695-3096-3/08 $25.00 © 2008 IEEE, DOI 10.1109/WAINA.2008.292, 1672

12

BER Analysis of 2x2 MIMO Spatial Multiplexing Under AWGN and Rician Channels For Different Modulations Techniques

Anuj Vadhera and Lavish Kansal

Lovely Professional University, Phagwara, Punjab, India

Abstract

Multiple-input–multiple-output (MIMO) wireless systems use multiple antennas at transmitting and receiving end to offer improved capacity and data rate over single antenna systems in multipath channels. In this paper we have investigated the Spatial Multiplexing technique of MIMO systems. Here different fading channels like AWGN and Rician are used for analysis purpose. Moreover we analyzed the technique using high level modulations (i.e. M-PSK for different values of M). Detection algorithms used are Zero-Forcing and Minimum mean square estimator. Performance is analyzed in terms of BER (bit error rate) vs. SNR (signal to noise ratio).

Keywords

Spatial Multiplexing (SM), Additive White Gaussian Noise (AWGN), Multiple Input Multiple Output (MIMO), Bit error rate (BER).

1. Introduction

Multiple antenna systems (MIMO) attract significant attention due to their ability of resolving the bottleneck of traffic capacity in wireless networks. MIMO systems are illustrated in Figure 1. The idea behind MIMO is that the signals on the transmitting (Tx) antennas and the receiving (Rx) antennas are combined in such a way that the quality (bit-error rate or BER) or the data rate (bits/sec) of the communication for each MIMO user will be improved. Such an advantage can be used to increase the network's quality of service. In this paper, we focus on the Spatial Multiplexing technique of MIMO systems.

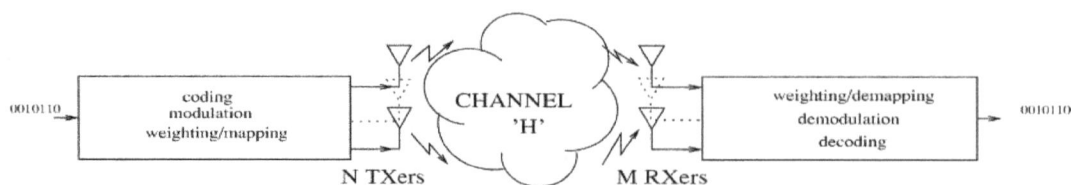

Figure 1.Diagram of MIMO wireless transmission system . Transmitter and receiver are equipped with multiple antennas

Spatial multiplexing is a transmission technique in MIMO wireless communication system to transmit independent and separately encoded data signals, called as streams, from each of the multiple transmit antennas. Therefore, the space dimension is reused or multiplexed more than one time. If the transmitter and receiver has N_t and N_r antennas respectively, the maximum spatial multiplexing order (the number of streams) is

$$N_S = \min(N_t, N_r) \tag{1}$$

The general concept of spatial multiplexing can be understood using MIMO antenna configuration. In spatial multiplexing, a high data rate signal is divided into multiple low rate data streams and each stream is transmitted from a different transmitting antenna. These signals arrive at the receiver antenna array with different spatial signatures, the receiver can separate these streams into parallel channels thus improving the capacity. Thus spatial multiplexing is a very powerful technique for increasing channel capacity at higher SNR values. The maximum number of spatial streams is limited by the lesser number of antennas at the transmitter or receiver side. Spatial multiplexing can be used with or without transmit channel knowledge.

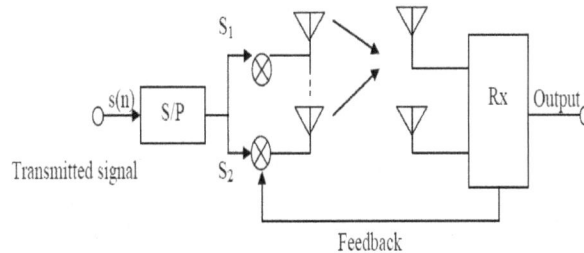

Figure 2.Spatial Multiplexing Concept

MIMO spatial multiplexing achieves high throughput by utilizing the multiple paths and effectively using them as additional channels to carry data such that receiver receives multiple data at the same time. The tenet in spatial multiplexing is to transmit different symbols from each antenna and the receiver discriminates these symbols by taking advantage of the fact that, due to spatial selectivity, each transmit antenna has a different spatial signature at the receiver . This allows an increased number of information symbols per MIMO symbol. In any case for MIMO spatial multiplexing, the number of receiving antennas must be equal to or greater than the number of transmit antennas such that data can be transmitted over different antennas. Therefore the space dimension is reused or multiplexed more than one time. The data streams can be separated by equalizers if the fading processes of the spatial channels are nearly independent. Spatial multiplexing requires no bandwidth expansion and provides additional data bandwidth in multipath radio scenarios [2].

2. MIMO SYSTEM

In MIMO system we use multiple antennas at transmitter and receiver side, they are extension of developments in antenna array communication. There are three categories of MIMO techniques. The first aims to improve the reliability by decreasing the fading through multiple spatial paths. Such technique includes STBC and STTC. The second class uses a layered approach to increase capacity. One popular example of such a system is V-BLAST suggested by Foschini et al. [2].

Finally, the third type exploits the knowledge of channel at the transmitter. It decomposes the channel coefficient matrix using SVD and uses these decomposed unitary matrices as pre- and post-filters at the transmitter and the receiver to achieve near capacity [3].

2.1. Benefits of MIMO system

MIMO channels provide a number of advantages over conventional Single Input Single Output (SISO) channels such as the array gain, the diversity gain, and the multiplexing gain. While the array and diversity gains are not exclusive of MIMO channels and also exist in single-input multiple-output (SIMO) and multiple-input single-output (MISO) channels, the multiplexing gain is a unique characteristic of MIMO channels. These gains are described in brief below:

2.2.1 Array Gain

Array gain is the average increase in the SNR at the receiver that arises from the coherent combining effect of multiple antennas at the receiver or transmitter or both. Basically, multiple antenna systems require perfect channel knowledge either at the transmitter or receiver or both to achieve this array gain.

2.2.2 Spatial Diversity Gain

Multipath fading is a significant problem in communications. In a fading channel, signal experiences fade (i.e they fluctuate in their strength) and we get faded signal at the receiver end. This gives rise to high BER. We resort to diversity to combat fading. This involves providing replicas of the transmitted signal over time, frequency, or space.

2.2.3 Spatial Multiplexing Gain

Spatial multiplexing offers a linear (in the number of transmit-receive antenna pairs or min (MR, MT) increase in the transmission rate for the same bandwidth and with no additional power expenditure. It is only possible in MIMO channels. Consider the cases of two transmit and two receive antennas. The stream is split into two half-rate bit streams, modulated and transmitted simultaneously from both the antennas. The receiver, having complete knowledge of the channel, recovers these individual bit streams and combines them so as to recover the original bit stream. Since the receiver has knowledge of the channel it provides receive diversity, but the system has no transmit diversity since the bit streams are completely different from each other in that they carry totally different data. Thus spatial multiplexing increases the transmission rates proportionally with the number of transmit-receive antenna pairs.

2.3 Modulation

Modulation is the process of mapping the digital information to analog form so it can be transmitted over the channel. Modulation of a signal changes binary bits into an analog waveform. Modulation can be done by changing the amplitude, phase, and frequency of a sinusoidal carrier. Every digital communication system has a modulator that performs this task. Similarly we have a demodulator at the receiver that performs inverse of modulation. There are several digital modulation techniques used for data transmission.

2.3.1 Phase Shift Keying

Phase-shift keying (PSK) is a digital modulation scheme that conveys data by modulating, the phase of a reference signal (the carrier wave). In M-ary PSK modulation, the amplitude of the transmitted signals is constrained to remain constant, thereby yielding a circular constellation. Modulation equation of M-PSK signal is:

$$s_i(t) = \sqrt{\frac{2E_S}{T}} \cos\left(2\pi f_c t + \frac{2\pi i}{M}\right) \qquad i=0,1....,M$$

(2)

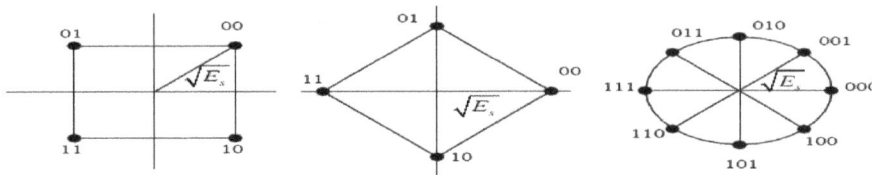

Figure 3.Constellation Diagrams of M-PSK
(a)   QPSK (b) QPSK (c) 8-PSK

2.4 Channels

Channel is transmission medium between transmitter and receiver. Channel can be wired or wireless. In wireless transmission we use air or space as medium and it is not as smooth as wired transmission since the received signal is not only coming directly from the transmitter, but the combination of reflected, diffracted, and scattered copies of the transmitted signal. These signals are called multipath components. AWGN and Rician channels are taken into consideration for the analysis.

2.4.1 AWGN Channel

AWGN channel is universal channel model for analyzing modulation schemes. In this model, a white Gaussian noise is added to the signal passing through it. Fading does not exist. The only distortion is introduced by the AWGN. AWGN channel is a theoretical channel used for analysis purpose only. The received signal is simplified to:

$$y(t)=x(t)+n(t) \qquad\qquad (3)$$

where n(t) is the additive white Gaussian noise.
   y(t) is the received signal
   x(t) is the input signal

2.4.2 Rician Channel

The direct path component is the strongest component that goes into deep fades compared to multipath components when there is line of sight. Such signal is approximated with the help of Rician distribution. The received signal can be simplified to:

$$y(t)=x(t)*h(t)+n(t) \qquad (4)$$

where h(t) is the random channel matrix having Rician distribution and n(t) is the additive white Gaussian noise.

The Rician distribution is given by:

$$P(r) = \frac{r^2}{\sigma^2} e^{\left(-\frac{r^2+A^2}{\sigma^2}\right)} I_0\left(\frac{A_r}{\sigma^2}\right) \quad \text{for } (A \geq 0, r \geq 0)$$

$$(5)$$

where A denotes the peak amplitude of the dominant signal and $I_0[.]$ is the modified Bessel function of the first kind and zero-order.

2.5 Detection Techniques

There are numerous detection techniques available with combination of linear and non-linear detectors. The most common detection techniques are ZF, MMSE and ML detection technique. The generalized block diagram of MIMO detection technique is shown in Figure 4.

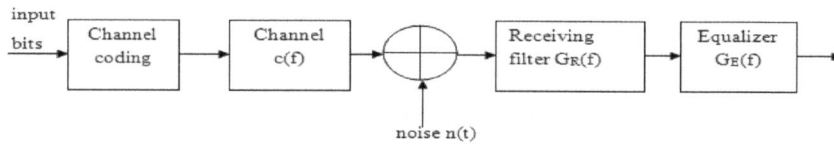

Fig. 4 Block Diagram of system with equalizer

2.5.1 Zero Forcing (ZF) Detection

The ZF is a linear estimation technique, which inverse the frequency response of received signal, the inverse is taken for the restoration of signal after the channel. The estimation of strongest transmitted signal is obtained by nulling out the weaker transmit signal. The strongest signal has been subtracted from received signal and proceeds to decode strong signal from the remaining transmitted signal. ZF equalizer ignores the additive noise and may significantly amplify noise for channel.

The basic Zero force equalizer of 2x2 MIMO channel can be modelled by taking received signal y_1 during first slot at receiver antenna as:

$$y_1 = h_{1,1}x_1 + h_{1,2}x_2 + n_1 = \begin{bmatrix} h_{1,1} & h_{1,2} \end{bmatrix}\begin{bmatrix} x_1 \\ x_2 \end{bmatrix} + n_1 \qquad (6)$$

The received signal y_2 at the second slot receiver antenna is:

$$y_2 = h_{2,1}x_1 + h_{2,2}x_2 + n_2 = \begin{bmatrix} h_{2,1} & h_{2,2} \end{bmatrix}\begin{bmatrix} x_1 \\ x_2 \end{bmatrix} + n_2 \qquad (7)$$

Where i=1, 2 in x_i is the transmitted symbol and i=1, 2 in $h_{i,j}$ is correlated matrix of fading channel, with j represented transmitted antenna and i represented receiver antenna, is the noise of first and second receiver antenna. The ZF equalizer is given by:

$$W_{ZF} = \left(H^H\right)^{-1} H^H \tag{8}$$

Where W_{ZF} is equalization matrix and H is a channel matrix. Assuming $M_R \geq M_T$ and H has full rank, the result of ZF equalization before quantization is written

as: $$y_{ZF} = \left(H^H H\right)^{-1} H^H y \tag{9}$$

2.5.2. Minimum Mean Square Estimator (MMSE)

Minimum mean square error equalizer minimizes the mean –square error between the output of the equalizer and the transmitted symbol, which is a stochastic gradient algorithm with low complexity. Most of the finite tap equalizers are designed to minimize the mean square error performance metric but MMSE directly minimizes the bit error rate. The channel model for MMSE is same as ZF [13],[14]. The MMSE equalization is

$$W_{MMSE} = \arg_G^{\min} E_{x,n}[\|x - x^\wedge\|^2] \tag{10}$$

Where is W_{MMSE} equalization matrix, H channel correlated matrix and n is channel noise

$$y_{MMSE} = H^H(HH^H + n_o I_n)^{-1} y \tag{11}$$

3. Results and Discussions

This paper analyzes the Spatial Multiplexing(SM) technique for 2x2 antenna configuration under different modulation techniques for different fading channels i.e. AWGN and Rician channels. Results are shown in the term of BER vs SNR plots.

3.1 Using ZF detection

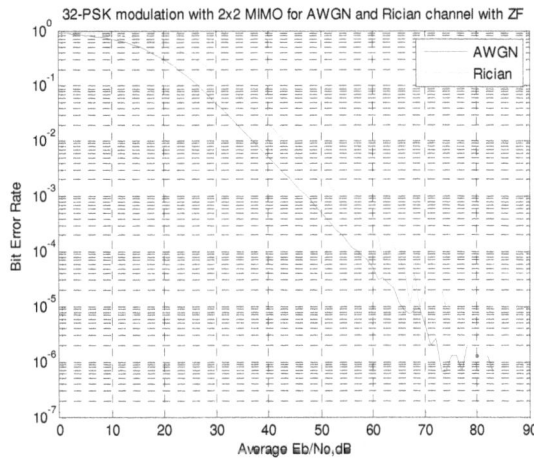

Figure 5(a).

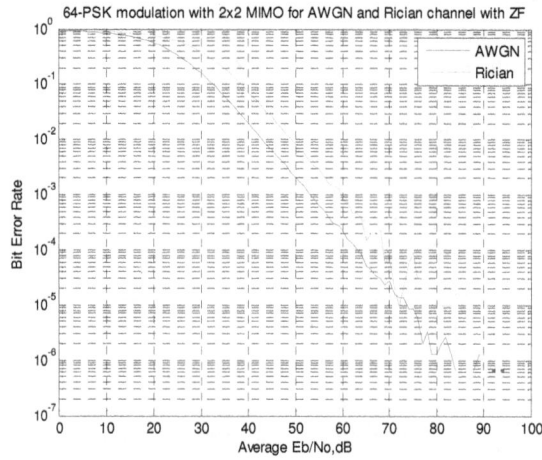

Figure 5(b).

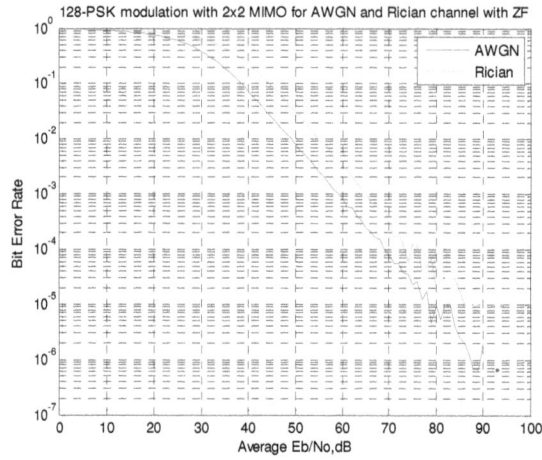

Figure 5(c).

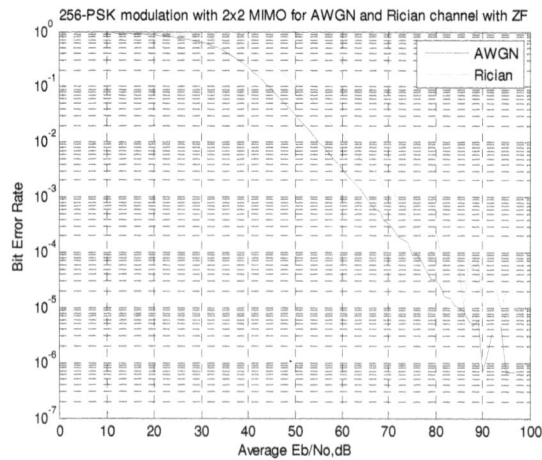

Figure 5(d).

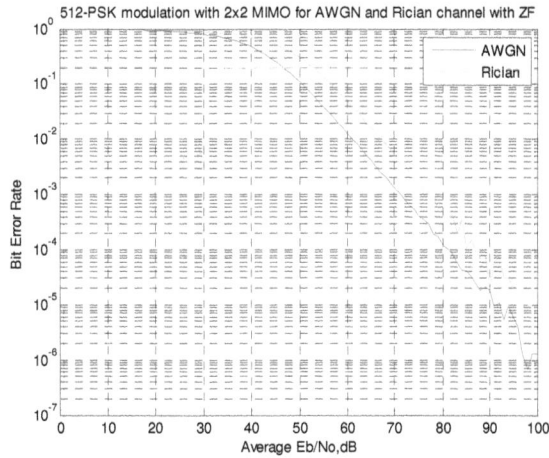

Figure 5(e).

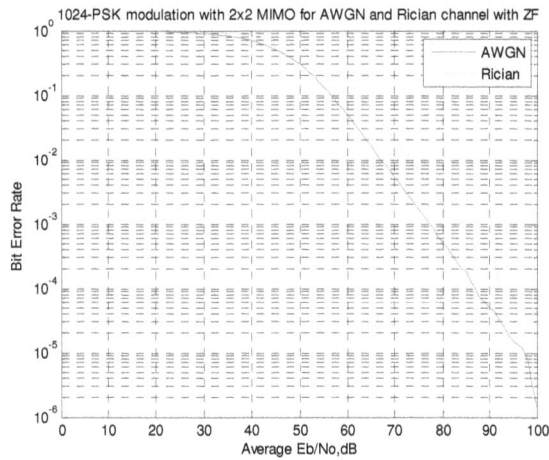

Figure 5(f).

Figure 5. BER vs. SNR plots over AWGN & Rician channel for SM technique using 2x2 MIMO System using ZF Equalization

a)32 PSK b) 64 PSK c) 128 PSK d) 256 PSK e) 512 PSK
f) 1024 PSK

Table 1. Comparison of different Modulation Techniques for Rician& AWGN Channel for 2x2 MIMO Spatial Multiplexing using ZF Equalization

Modulations	Rician channel	AWGN channel	Improvement
32-PSK	62dB	57dB	5dB
64-PSK	63dB	69db	6dB

128-PSK	74dB	69dB	5dB
256-PSK	81dB	75dB	6dB
512-PSK	86dB	81dB	5dB
1024-PSK	93dB	87dB	6dB

From table we depict that at 32-PSK, 128-PSK, 512-PSK there is difference of 5dB between channels and there is difference of 6dB at 64-PSK, 128-PSK and 1024-PSK at BER of 10^{-4}. Table shows the improvement in terms of decibels shown by proposed system employing SM technique for 2x2 MIMO system for different modulation schemes over different channels.

3.2 Using MMSE detection

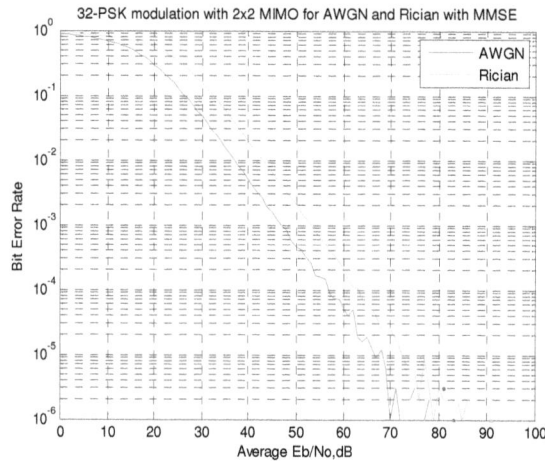

Figure 6(a).

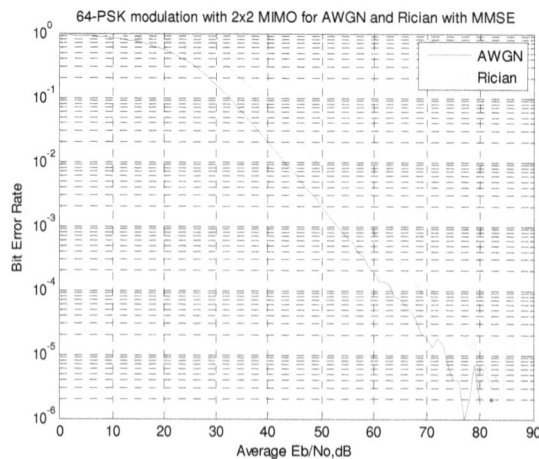

Figure 6(b).

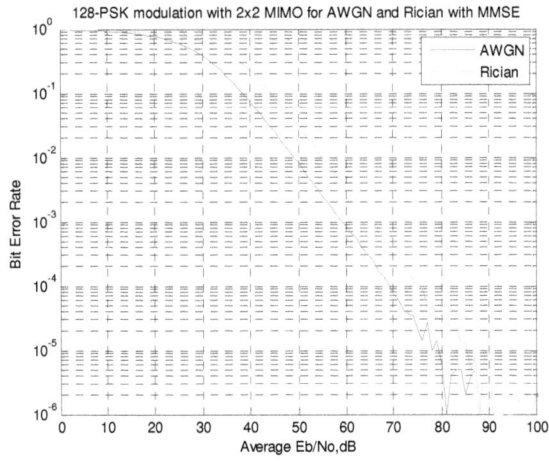

Figure 6(c).

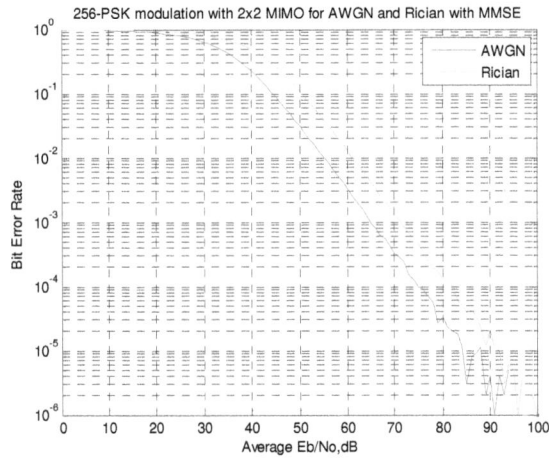

Figure 6(d).

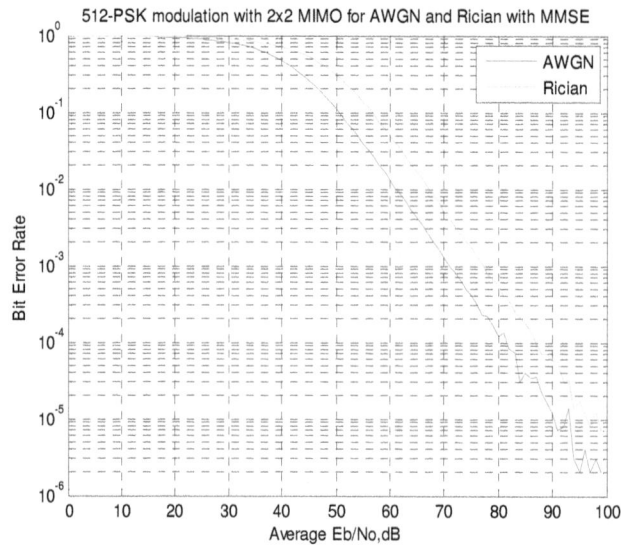

Figure 6(e)

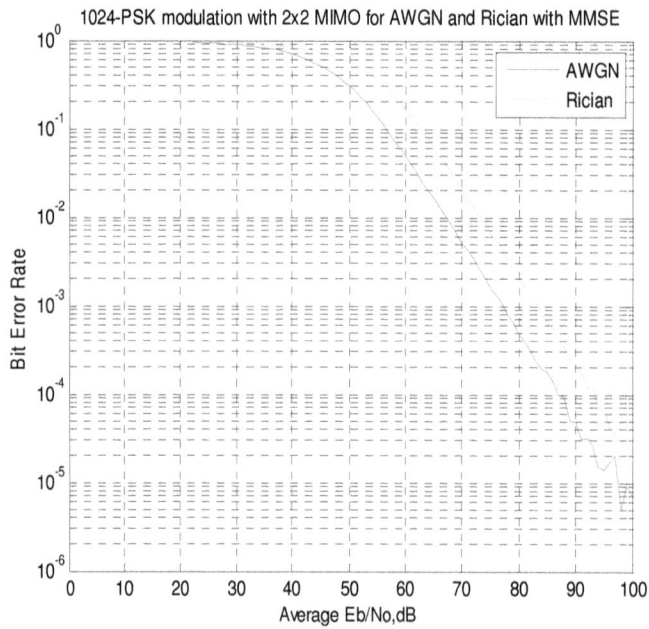

Figure 6(f)

Fig. 6 BER vs. SNR plots over AWGN & Rician channel for SM technique using 3x3 MIMO using MMSE Equalization
a) 32 PSK b) 64 PSK c) 128 PSK d) 256 PSK e) 512 PSK f) 1024 PSK

Table 2. Comparison of different Modulation Techniques for Rician &AWGN Channel for 2x2 MIMO Spatial Multiplexing using MMSE Equalization

Modulations	Rician channel	AWGN channel	Improvement
32-PSK	63dB	57dB	6dB
64-PSK	70dB	63dB	7dB
128-PSK	75dB	69dB	6dB
256-PSK	82dB	76dB	6dB
512-PSK	86dB	82dB	4dB
1024-PSK	93dB	87dB	6dB

It can be seen from table that at 32-PSK, 128-PSK, 256-PSK and 1024-PSK there is an improvement of 6dB. At 64-PSK and 512-PSK there is difference of 7dB and 4dB at BER of 10^{-4}. Table shows the improvement in terms of decibels shown by proposed system employing SM technique for 2x2 MIMO system for different modulation schemes over different channels.

4. CONCLUSIONS

In this paper, an idea about the performance of the MIMO-SM technique at higher modulation levels is presented. We implemented 2x2 antenna configuration and used different signal detection technique at receiver end. It can be concluded BER is greater in Rician channel as compared to AWGN channel.

Also BER (bit error rate) increases as the order of the modulation order i.e. M increases. This increase is due to the fact that as the value of M increases distances between constellation points decreases which in turn makes the detection of the signal corresponding to the constellation point much tougher The solution to this problem is to increase the value of the SNR so, that the effect of the distortions introduced by the channel will also goes on decreasing, as a result of this, the BER will also decreases at higher values of the SNR for high order modulations.

ACKNOWLEDGEMENTS

I express my sincere thanks to my esteemed and worthy guide Mr. Lavish Kansal, Assistant Professor, Electronics and Communication Engineering Department, Lovely Professional University, Phagwara, for his valuable advice, motivation, guidance, encouragement, efforts, timely help and the attitude with which he solved all of my queries regarding thesis work. I am highly grateful to my entire family and friends for their inspiration and ever encouraging moral support, which enable me to pursue my studies.

REFERENCES

[1] H. Jiang and P. A. Wilford, "A hierarchical modulation for upgrading digital broadcasting systems," IEEE Transaction on Broadcasting, vol. 51, pp. 222-229, June 2005.

[2] P. W. Wolniansky, G. J. Foschini, G. D. Golden and R. A.Valenzuela, "V-BLAST: an architecture for realizing very high data rates over the rich- scattering wireless channel," In Proceeding of International symposium on Signals, Systems Electronics, pp. 259-300, October 1998.

[3] J. Ha, A. N. Mody, J. H. Sung, J. Barry, S. Mclaughlin and G. L. Stuber, "LDPC coded OFDM with Alamouti/SVD diversity technique," IEEE Journal on Wireless Personal Communication, Vol. 23 , Issue 1,pp. 183-194,Oct. 2002.

[4] P. S. Mundra , T. L. Singal and R. Kapur, "The Choice of A Digital Modulation ,Schemes in A Mobile Radio System", In proceedings of IEEE Vehicular Technology Conference, Issue 5, pp 1-4,(Secaucus, NJ)1993.

[5] P. Liu & I1-Min Kim, "Exact and Closed-Form Error Performance Analysis for Hard MMSE-SIC Detection in MIMO Systems", IEEE Transactions on Communication, Vol. 59, no. 9, September 2011.

[6] P. Sanghoi & L. Kansal," Analysis of WiMAX Physical Layer Using Spatial Multiplexing Under Different Fading Channels", SPIJ, Vol.(6),Issue(3),2012.

[7] C. Wang & E. K. S. Au, R. D Murch, W. H. Mow & V. Lau," On the Performance of the MIMO Zero-Forcing Receiver in the Presence of Channel Estimation Error", IEEE Transactions on Wireless Communication, Vol. 6,no.3,2007.

[8] X. Zhang, Y. Su & G. Tao, "Signal Detection Technology Research of MIMO-OFDM System", 3rd International Congress on Image and Signal Processing, pp 3031-3034, 2010.

[9] I. Ammu & R. Deepa, "Performance Analysis of Decoding Algorithms in multiple antenna systems", IEEE, pp 258-262, 2011.

[10] H. B. Voelcker, "Phase-shift keying in fading channels" , In IEEE Proceeding on Electronics and Communication Engineering, Vol. 107, Issue 31, pp 31-38, 1960.

[11] D. S. Shiu, G. J. Foschini, M. J. Gans, and J. M. Kahn, "Fading correlation and its effect on the capacity of multi-element antenna systems", IEEE Transaction on Communication, Vol. 48, pp. 502–513, 2000.

[12] G. J. Foschini, K. Karakayali, and R. A.Valenzuela, "Coordinating multiple antenna cellular networks to achieve enormous spectral efficiency," Communications, IEEE Proceedings, Vol. 153, pp. 548–555, 2006.

[13] J. S. Thompson, B. Mulgrew and Peter M. Grant, "A comparison of the MMSE detector and its BLAST versions for MIMO channels", IET seminar on Communication System from Concept to Implementation, pp. 1911-1916, 2001.

[14] X. Zhang and Sun-Yuan Kung, "Capacity analysis for parallel and sequential MIMO equalizers", IEEE Transaction on Signal processing, Vol. 51, pp. 2989-3002, 2003.

13

ADVANCED ANTENNA TECHNIQUES AND HIGH ORDER SECTORIZATION WITH NOVEL NETWORK TESSELLATION FOR ENHANCING MACRO CELL CAPACITY IN DC-HSDPA NETWORK

Muhammad Usman Sheikh[1], Jukka Lempiainen[1] and Hans Ahnlund[2]

[1]Department of Communications Engineering, Tampere University of Technology
[2]European Communications Engineering Ltd, Tekniikantie 12, Espoo Finland

ABSTRACT

Mobile operators commonly use macro cells with traditional wide beam antennas for wider coverage in the cell, but future capacity demands cannot be achieved by using them only. It is required to achieve maximum practical capacity from macro cells by employing higher order sectorization and by utilizing all possible antenna solutions including smart antennas. This paper presents enhanced tessellation for 6-sector sites and proposes novel layout for 12-sector sites. The main target of this paper is to compare the performance of conventional wide beam antenna, switched beam smart antenna, adaptive beam antenna and different network layouts in terms of offering better received signal quality and user throughput. Splitting macro cell into smaller micro or pico cells can improve the capacity of network, but this paper highlights the importance of higher order sectorization and advance antenna techniques to attain high Signal to Interference plus Noise Ratio (SINR), along with improved network capacity. Monte Carlo simulations at system level were done for Dual Cell High Speed Downlink Packet Access (DC-HSDPA) technology with multiple (five) users per Transmission Time Interval (TTI) at different Intersite Distance (ISD). The obtained results validate and estimate the gain of using smart antennas and higher order sectorization with proposed network layout.

KEYWORDS

System capacity, Sectorization, Advanced antenna techniques, Switched beam antenna, Adaptive antenna, Dual Cell High Speed Downlink Packet Access.

1. INTRODUCTION

Rising trend of packed switched traffic and high capacity requirement in mobile networks have urged the researchers to think about new antenna designs and possible network layouts for future cellular networks. Current and future capacity demands of next generation mobile networks cannot be achieved by using traditional macro cells only. It has been noted several times that macro cells are not able to offer high data rates homogeneously over the entire cell area, and most of the network capacity is lost due to interference coming from the neighbor cells. The increasing demand of new advanced mobile services with different Quality of Service (QoS) requirement in cellular systems has led to the development and evolution of new technologies. Concepts of micro cells and femto cells have been proposed to improve the system capacity in high density traffic areas [2]. However, to reduce the fixed costs such as electricity, transmission, rentals etc., adding new cells and sites should be avoided. Maximum capacity utilization of macro cells should be guaranteed by adopting new network tessellation and by employing possible advanced antenna solutions, including smart antennas. Smart antennas have gained enormous popularity in the last few years, and have been able to grab the attention for its ability to improve the performance of cellular systems [3].

Moreover, cell and system capacities are related to network layout, antennas deployment techniques, orientation and beamwidth of antennas. Directional antennas with optimum electrical or mechanical tilt are used to get required coverage with minimum interference [2]. Antenna configuration i.e. antenna height, azimuth, radiation pattern and beamwidth has deep impact on cell capacity [4– 6]. Different network tessellations have been compared in [7], and it was noted that for 3-sector sites, cloverleaf layout offers the lowest interference level, and thus should have the best cell and system capacity for macro cells. Thus, cloverleaf is a good basis for nominal planning of mobile networks with 3-sector sites. However, for higher order of sectorization, cloverleaf layout cannot be used and new tessellation is needed to combat the problem of interference. Base station antenna configuration needs to be optimized to attain minimum inter cell interference [1], [8]. The conventional cellular concept approach uses fixed beam position with wide beamwidth. Whereas, advanced approach of smart antenna employs multiple narrow beams and beam steering for each user in a cell. Adaptive algorithms form the heart of antenna array processing network. The processor based on different beamforming algorithms does the complex computation for beamforming [9]. Achieved user SINR and user throughput strongly relies on interference management and inter-cell interference avoidance [10]. Handovers between cells due to mobility of user, and software features have their own impact on cell capacity. However, this research work does not deal with these issues.

Over the last decade, services like multimedia messaging, video streaming, video telephony, positioning services and interactive gaming have become an integral part of everyday life. These services are the driving force in reshaping the cellular technologies. Universal Mobile Telecommunication System (UMTS) has been the most popular choice for 3G mobile communication systems, but UMTS had challenges in meeting the requirement of high data rate services. High Speed Downlink Packet Access (HSDPA) was for first time introduced in Release 5 of 3GPP specifications [8], [10]. The evolution of HSDPA continued and later in Release 8, the concept of Dual Cell HSDPA was floated in which the radio resources of two adjacent HSDPA carriers were aggregated with the help of joint scheduler. The main target of DC-HSDPA was not only to improve the user's throughput in the close vicinity of base station rather it also enhances the user's throughput homogeneously over whole cell area. DC-HSDPA offers theoretical peak data rate of 42 Mbps, improved spectral efficiency, and enhanced user experience with low delays or latency [8], [11].

In this paper user's SINR value, average SINR over the cell, mean cell throughput, mean site throughput, user's throughput and user's probability of no data transfer will be taken as merits of performance. Statistical analysis with 10th, 50th, 90th percentile, and mean value is also presented in this paper. The rest of the paper is organized as follows. Section II deals with theoretical aspects of cell capacity. Section III explains different antenna techniques. Description of simulation tool and environment, simulated cases, and simulation parameters is presented in section IV. Simulation results and their analysis are given in section V. Finally, section six concludes the paper.

2. CELLULAR THEORY

2.1. Interference and Cell Capacity

Theoretical maximum cell capacity can be estimated by Shannon capacity equation (bits/s) for Additive White Gaussian Noise (AWGN) channel as given in equation (1), [1], [4]

$$C = W log_2 \left(1 + \frac{S}{N}\right) = W log_2 \left(1 + \frac{E_b R}{N_0 W}\right) \qquad (1)$$

where W is the bandwidth available for communication, S is the received signal power which can be denoted as energy per information bit E_b, multiplied with the information rate R. A variable N is the noise power impairing the received signal. The noise power can be defined as noise spectral density N_0 multiplied with the transmission bandwidth W. Signal to Noise ratio (S/N) can be extended to Signal to Interference plus Noise Ratio (SINR) by including interference from own cell and also co-channel interference coming from neighbor cells. HSDPA is a WCDMA based network, and the total interference is a sum of three interference sources; own (serving) sector signals, other site/sector signals, and thermal noise. In downlink direction, the total interference I_{DL} for any particular user at a given location is given by equation (2), [8]

$$I_{DL} = I_{other} + I_{own} + P_N \qquad (2)$$

$$I_{other} = \sum_{i=1}^{k} \frac{Pt_i}{L_i} \qquad (3)$$

$$I_{own} = \frac{(Pt_S - S_j)}{L_S} * (1 - \alpha) \qquad (4)$$

In equation (2), P_N is a thermal noise power. In equation (3), I_{other} is the total received power from other sectors of the network, and is a sum of other cells interfering sources. Pt_i is a total transmit power and L_i is a path loss for i^{th} neighbor cell. In equation (4), I_{own} is the total received interference from own sector, where Pt_S and L_S are the total transmit power and path loss respectively of serving cell. Where S_j is the received power of HS-PDSCH of the jth user from the serving NodeB. α denotes orthogonality factor. Orthogonality is a measure for level of interference caused by own sector signals. For perfectly orthogonal DL channelization codes α is equals to 1. In HSDPA technology, code orthogonality is partly lost ($\alpha < 1$) in wireless radio environment due to multipath propagation [8], [12]. The ratio of I_{other}/I_{own} is a commonly used measure of sector overlap and interference in the network layout. The SINR represents the quality of the received signal. In the downlink direction the receiver input, SINR is defined as

$$SINR_{DL} = \frac{S_j}{I_{other} + I_{own} + N} \qquad (5)$$

2.2. Network layouts and inter cell interference

In initial nominal plan for mobile network, regular network layouts are used for guidance on selection of nominal site location, order of sectorization, and azimuth direction. There is triangular, square, and hexagonal tessellation for 3-sector site, but the most commonly used tessellation is cloverleaf layout as shown in [7]. These tessellations are chose to form continuous coverage of the mobile network. In Fig1a, cloverleaf layout is shown, that is formed by using hexagonal geometry of cell. In cloverleaf layout, all the interfering sites of the first tier of interferer are pointing at the null of serving site. Authors of this paper propose a name "Snow Flake" layout for the enhanced cellular network tessellation for six sector site presented in [13]. Snow flake tessellation is shown in Fig1b. This paper presents a novel network layout for 12-sector site, as shown in Fig.1c and calls it as "Flower" layout. SINR calculations include own cell and neighbor cell interference as given in equation (5). These interferences are related to propagation loss i.e. path loss L_S and L_i between serving NodeB and interfering NodeBs respectively. Especially inter cell interference depends heavily on chosen network layout i.e. how base stations are deployed in a network, antenna configuration, and azimuth. Network layout has significant impact on interference management and hence on capacity of macrocellular network. One way to compare different network layouts or different antenna configurations is to compare the interference coming from neighbor cells to serving cells. It has been shown in [7] that for 3-sector sites, cloverleaf is the most defensive for interference and thus provides high capacity gain. In this article, for 3-sector sites only cloverleaf layout is considered for network simulations.

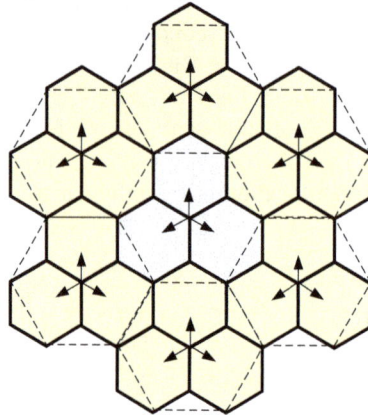

Fig.1. (a) 3-sector "Cloverleaf" layout

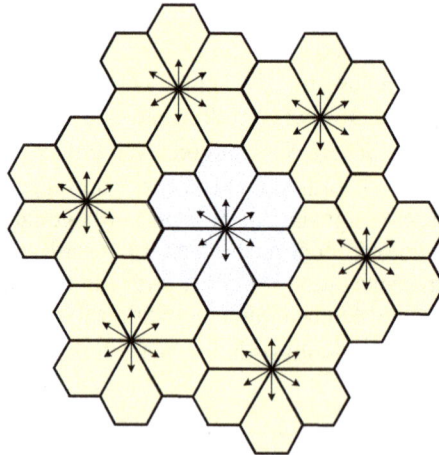

Fig.1. (b) 6-sector "Snow Flake" layout

Fig.1. (c) 12-sector "Flower" layout

3. ANTENNA THEORY

The functionality of antenna depends on number of factors including physical size of an antenna, impedance (radiation resistance), beam shape, beam width, directivity or gain, polarization etc [14]. By definition, an antenna array consists of more than one antenna element. The radiation pattern of an antenna array depends on number of antenna elements used in array. The more elements there are, the narrower beam can be formed. Planar arrays are capable of making a narrow beam in horizontal as well as in vertical plane. Therefore, planar array beams are also called "Pencil Beams". Smart antennas with ability of beam steering can be constructed by adding "Intelligence" to planar arrays. Smart antenna improves the coverage of cell by concentrating more power in a narrow beam, enhances the cell capacity and offers increased data rates by offering high signal to interference plus noise ratio [15]. By avoiding interference and increasing signal power, smart antenna improves link quality and helps in combating large delay dispersion [16].

In the research work of this paper, three different type of antenna were taken into account i.e.1) Conventional 65^0 wide beam antenna, 2) Switched beam smart antenna and 3) Full adaptive beam antenna.

3.1. Conventional wide beam antenna

In traditional cellular networks, three-sectored approach with 65^0 wide beam antenna has been in used for long time due to lower interference compared to 120^0 wide beam antenna. To further improve the performance of fixed wide beam antenna, electrical or mechanical tilting can be used [6]. Base station antennas can be dropped down to building walls but then the propagation environment is not any more related to macro cells, rather shifts to micro cell environment. Other possibility is to modify and narrow the radiation pattern with the help of antenna arrays.

3.2. Switched beam smart antenna

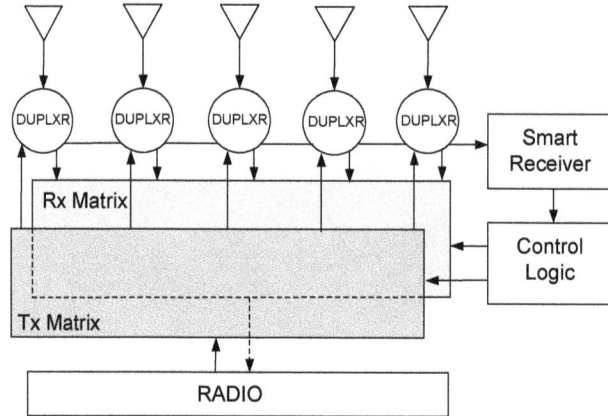

Fig.2. Block diagram of switched beam smart antenna system

Switched beam antenna approach is the extension of conventional cellular sectorisation method, in which single 120^0 wide macro sector is divided into several micro sectors. A switched beam antenna is a combination of multiple narrow beams in predetermined directions, overlapping over each other. It covers the desired cell area with finite number of narrow fixed beams, where each beam can serve a single user or multiple users [3]. Switched beam antenna does not steer or adapt the beam with respect to the desired signal. In this type of antenna, a RF switch connected to fixed beams controls the beam selection based on the beam-switching algorithm. A switch selects the "Optimum" beam to provide service to mobile station. The optimum beam here refers to the beam that offers the highest SINR value. In some cases, maximum received power for the user can be used as a beam selection criterion. During user mobility, switched beam antenna tracks the user and continuously updates the beam selection to ensure high quality of service [17]. The general block diagram of switched beam smart antenna system is shown in Fig.2 [18].

It consists of an array of antennas that divides the macro sector into several micro sectors. A precise switched beam antenna can be implemented by using "Butler Matrices" [16], [18]. It uses a smart receiver for detecting and monitoring the received signal power for each user at each antenna port. Based on the measurement made by the smart receiver and beam selection algorithm, the control logic block determines the most favorable beam for specific user. The RF switch part governed by the control logic (brain of switched beam antenna) activates the path from the selected antenna port to the radio transceiver. Switched beam antenna offers higher directivity with less interference and thus provides gain over conventional antenna. Theoretically, gain of using switched beam antenna over conventional wide single beam antenna is directly proportional to the number of beams. For a given sector containing U beams, resultant increased gain is given by equation (6) [18]. Switched beam approach is simpler and easier to implement compared to fully adaptive beam approach.

$Gain = 10\text{Log}(U)$ (6)

An example of the horizontal radiation pattern of 65^0, 32^0, and 16^0 HPBW antenna is depicted in Fig.3a, 3b and 3c respectively. Radiation pattern of seven switched beam antenna with each beam of 8^0 HPBW is shown in Fig.3d.

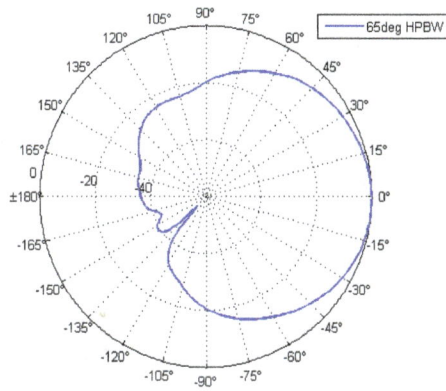

Fig.3. (a) Radiation pattern of conventional 65^0 beamwidth antenna used in 3-sectored site

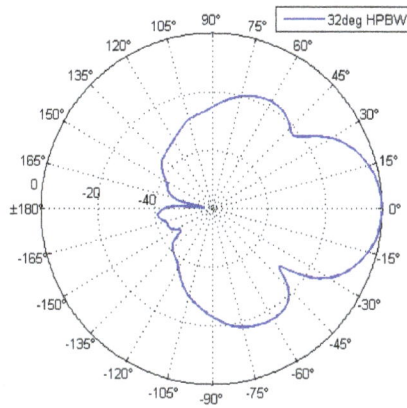

Fig.3. (b) Radiation pattern of narrow 32^0 beamwidth antenna used in 6-sectored site

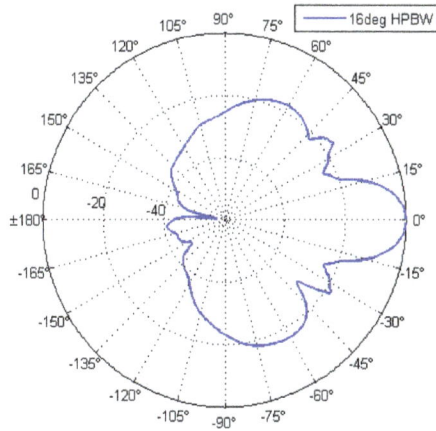

Fig.3. (c) Radiation pattern of narrow 16^0 beamwidth antenna used in 12-sectored site

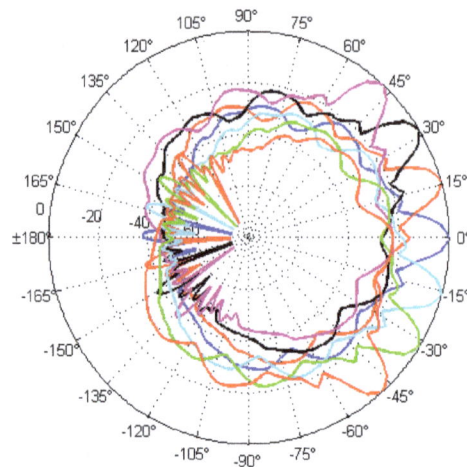

Fig.3. (d) Radiation pattern of switched beam antenna with seven beams of 8^0 HPBW

3.3. Full adaptive beam antenna

Adaptive antenna exploits the array of antenna elements to achieve maximum gain in desired direction while rejecting interference coming from other directions. Adaptive antennas are more complex than multi beam switched antennas. While butler matrices are operating on the RF domain, adaptive antennas use a linear combination of signals, and process them in the baseband. Adaptive antenna can steer its maxima and nulls of the array pattern in nearly any direction in response to the changing environment [16]. The basic idea behind adaptive antenna is the same as in switched beam antenna i.e. to maximize the SINR values. While the multiple switched beam antennas have a limited selection of directions to choose the best beam, an adaptive antenna can freely steer its beam in correspondence to the location of user. Smart antenna employs Direction of Arrival (DOA) algorithm to track the signal received from the user, and places nulls in the direction of interfering users and maxima in the direction of desired user [19]. On the other hand, since adaptive antennas needs more signal processing, multiple switched beam antennas are easier to implement and have the advantage of being simpler, and less expensive compared to adaptive antennas. The overall capacity gain of smart antennas is expected to be in the range of 100% to 200%, when compared with conventional antennas [3].

Beam forming algorithms used in adaptive antennas are generally divided into two classes with respect to the usage of training signal i) Blind Adaptive algorithm and ii) Non-Blind Adaptive algorithm [20]. In a non-blind adaptive beam forming algorithm, a known training signal d(t) is sent from transmitter to receiver during the training period. The beamformer uses the information of the training signal to update its complex weight factor. Blind algorithms do not require any reference signal to update its weight vector; rather it uses some of the known properties of desired signal to manipulate the weight vector. Fig.4 shows the generic beam forming system based on non-blind adaptive algorithm, which requires a training (reference) signal [19].

The output of the beamformer at time n, $y(n)$, is given by a linear combination of the data at the k antenna elements. The baseband received signal at each antenna element is multiplied with the weighting factor which adjusts the phase and amplitude of the incoming signal accordingly. The sum of this weighted signal results in the array output $y(n)$. On the basis of adaptive algorithms, entries of weight vector w are adjusted to minimize the error $e(n)$ between the training signal $d(n)$ and the array output $y(n)$. The output of the beamformer $y(n)$ can be expressed as given in equation (7), [20]

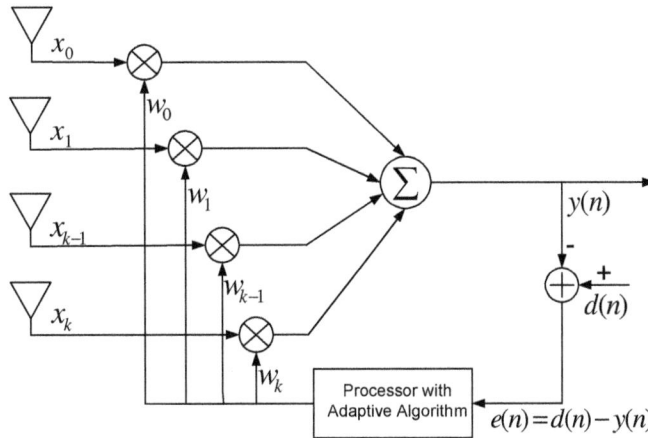

Fig.4. Block diagram of adaptive beamforming system

$y(n) = \boldsymbol{w}^H(n)\boldsymbol{x}(n)$ \qquad (7)

$\boldsymbol{w}(n) = [w_1(n) \ w_2(n) \dots \dots w_{k-1}(n) \ w_k(n)]$ \qquad (8)

$\boldsymbol{x}(n) = [x_1(n) \ x_2(n) \dots \dots x_{k-1}(n) \ x_k(n)]$ \qquad (9)

$e(n) = d(n) - y(n)$ \qquad (10)

where $\boldsymbol{w}(n)$ is the weight vector with $w_k(n)$ a complex weight for kth antenna element at time instant n, and $[.]^H$ denotes Hermitian (complex conjugate) transpose. $x_k(n)$ is the received baseband signal at kth antenna element [9], [20]. Least Mean Square (LMS), Normalized Least Mean Square (NLMS), Recursive Least Squares (RLS), and Direct Matrix Inversion (DMI) are examples of non-blind adaptive algorithm, whereas Constant Modulus Algorithm (CMA) and Decision Directed (DD) algorithms are examples of blind adaptive algorithm [9], [19-20]. These beamforming algorithms have their own pros and cons as far as their computational complexity, convergence speed, stability, robustness against implementation errors and other aspects are concerned.

4. SYSTEM SIMULATIONS

4.1. Simulation Environment

MATLAB was used as a simulation tool for carrying a campaign of simulations. Monte Carlo type of simulation was done with 5000 iterations with multiple users. It was aimed to model a network as realistic as possible. All system simulations for three sectored sites were done with macro cell cloverleaf layout. Snow flake and flower tessellation was selected for 6-sector 12-sector sites respectively. Base station grid of 19 sites was built, where single middle site in the middle has six sites in the first tier of interferer, and 12 sites in the second tier of interferer. All the interfering sites were at equal Intersite Distance (ISD) as shown in Fig.5(a,b,c), with same site parameters. Base station antenna height was set to 25m, which is typical value in city centre areas where 5-7 floor buildings exists. Power required for common pilot channel and signaling was taken into account. Frequency band of 2100MHz was used in simulations because DC-HSDPA system was selected as an example technology. Simulations were done with flat terrain, and Okumura-Hata model was used for calculating path loss between user and NodeB. Fading component is modelled with log normal distribution having zero mean and 5dB of standard deviation. Orthogonality factor used in equation (4) for computing own cell interference follows

Gaussian curve with maximum of 0.97 at site location and 0.7 at cell edge.

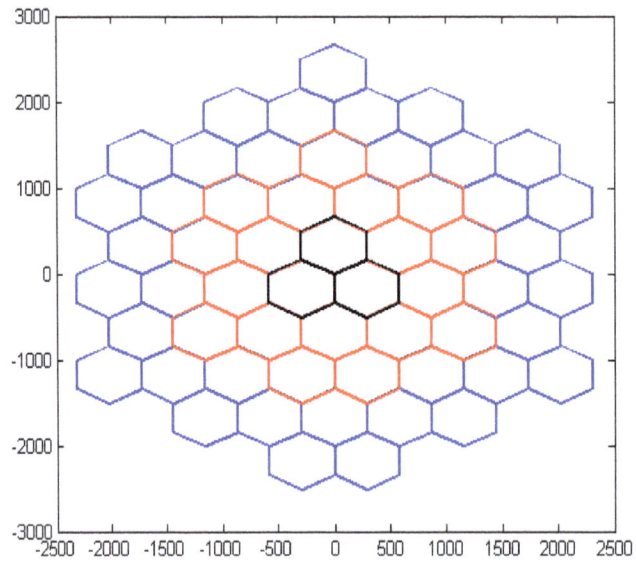

Fig.5. (a) Grid of nineteen 3-sector sites used in simulation with clove-leaf topology

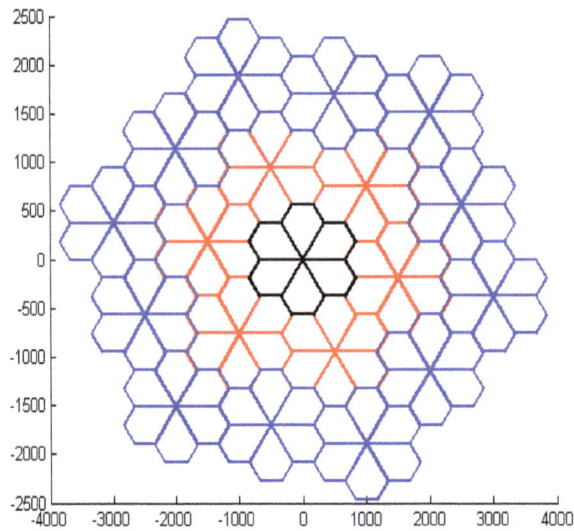

Fig.5. (b) Grid of nineteen 6-sector sites used in simulation with snow flake topology

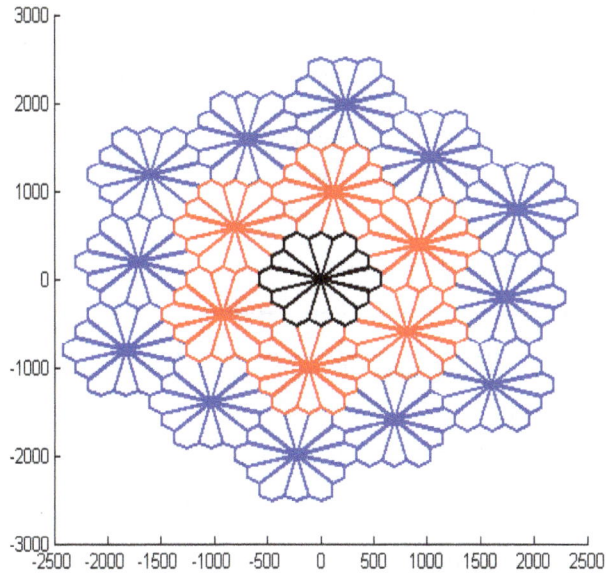

Fig.5. (c) Grid of nineteen 12-sector sites used in simulation with flower topology

4.2. Simulation cases and simulation procedure

Following three cases were considered for simulations.

- **3 Sector:** It is the most common scenario in which each site has three sectors and every sector has single 65^0 half power beamwidth antenna. This acts a reference case for comparing with higher order sectorization and advanced antenna case. Fig.3a shows the radiation pattern of an antenna used for simulations, with no electrical or mechanical tilt, and with maximum antenna gain of 15.39dB.
- **6 Sector:** It is the case in which each site has six sectors, and every sector has single 32^0 half power beamwidth antenna. Fig.3b shows the radiation pattern of an antenna used for simulations, with no electrical or mechanical tilt, and with maximum antenna gain of 18.20dB.
- **12 Sector:** In this case, each site comprises of 12 narrow sectors, and every sector has 16^0 HPBW antenna. Fig.3c shows the radiation pattern of an antenna used for simulations, with no electrical or mechanical tilt, and with maximum antenna gain of 21.15dB.
- **7 Switched beams:** This case represents multiple fixed switched beam scenario, where single sector is covered by seven potential narrow beams. Each narrow beam has eight degree HPBW with a spacing of 16^0 between the beams as shown in Fig.3d. Only that beam which has smallest deviation angle with respect to its main beam for user becomes active for that particular user. No down tilting was assumed, and each beam has maximum antenna gain of 23.55dB.
- **Adaptive beam:** In this last scenario, adaptive antennas are used to form an accurate beam for each individual user. In this scenario, narrow beam of six degree in the horizontal plane is steered precisely to the serving user, keeping user in the middle of the beam for maximum gain. Adaptive antenna have maximum gain of 24.5dB.

Key parameters related to DC-HSDPA systems used in simulations are presented in Table I. For each iteration, 5 users with full traffic buffer in each cell were created. Users were homogenously spread over the whole cell area on the flat terrain. In this simulation, DC-HSDPA serves five code multiplexed users per Transmission Time Interval (TTI). Out of total 16 codes, maximum of 15 codes were available for High Speed Physical Downlink Shared Channel (HS-PDSCH). Total transmission power for HS-PDSCH and available codes were equally distributed among the five users in each TTI. In the serving cell to compute the received signal value, Okumura-Hata model was used to calculate the path loss between the user and serving NodeB. Simulator supports Adaptive Modulation and Coding (AMC), and in these simulations eight different Modulation and Coding Schemes (MCS) were considered with 64QAM 5/6 coding rate as highest and QPSK 1/2 coding rate as lowest possible MCS. As throughput is the function of SINR, hence later SINR information was employed to compute each user throughput. Cell throughput in each TTI is the sum of individual users' throughput. Post processing of data was done to get the results in refined form.

Table I. General DC-HSDPA simulation parameters

Parameter	Unit	Value
DC-HSDPA Downlink		
Users per TTI	No.	5
Operating frequency band	MHz	2100
Bandwidth	MHz	5 + 5
Chip rate	Mcps	3.84
Total HS-PDSCH Codes	No.	15
Max HS-PDSCH power	dBm	41.63
HS-SCCH power	dBm	26
Processing gain	dB	12.04
HSDPA loading	%	70
Interference margin	dB	5.2
UE noise figure	dB	8.0
Downlink activity factor		1.0

5. SIMULATION RESULTS AND ANALYSIS

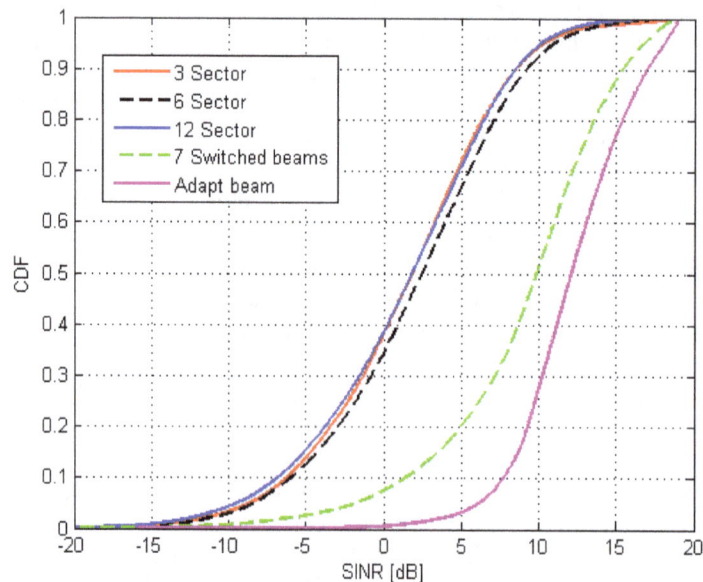

Fig.6. CDF plot of user SINR with 5 users per TTI at 1000m ISD

Fig.6 shows the Cumulative Distribution Function (CDF) of the user SINR with 5 users per TTI at 1000m ISD for different cases. Clearly switched beam antenna shows better performance in terms of offering higher SINR compared to 65^0, 32^0, and 16^0 wide beam antenna used in 3-sector, 6-sector and 12 sector sites respectively. But adaptive beam antenna outperforms and shows superior performance compared to all other cases. By analyzing the curves shown in Fig.6 it can be deduced that adaptive and switched beam antennas served the purpose of improving user experience by reducing the interference and enhancing the received SINR. The CDF curve of SINR for the case of adaptive beam is on the extreme right position, indicating that the SINR for the users is improved on average. It is also important to note that the average user SINR does not deteriorate by increasing the order of sectorization and almost similar performance is shown by 3-sector, 6 sector and 12-sector sites. However, 6-sector site offers slightly better performance compared to 3 and 12-sector sites. Adaptive beam antenna performed well in the close vicinity of the NodeB as well as near the cell edge, as 80% of the samples are concentrated in a narrow range of 9.12dB, starting from 7.72dB to 16.84dB of user SINR. But for other traditional antenna cases, eighty percent of SINR values has wide span and spread over the range of around 14.96dB, starting from -6.3 to 8.66dB. Statistical analysis of user SINR is presented in Table II.

Table II. Statistical Analysis of User SINR

	10 percentile user SINR (dB)	50 percentile user SINR (dB)	90 percentile user SINR (dB)	Mean user SINR (dB)	STD user SINR (dB)	Relative SINR gain (dB
3-Sector	-6.22	1.83	8.51	1.44	5.89	0
6-Sector	-5.99	2.44	9.22	1.98	5.96	0.54
12-Sector	-6.87	1.78	8.50	1.23	6.05	-0.21
7 Switched beam	1.36	9.83	15.41	8.94	5.72	7.50
Adaptive beam	7.72	12.10	16.97	12.07	3.73	10.63

Relative SINR gain shown in Table II is the relative gain in dB with respect to the mean SINR value of 3-sector case. It was learned that adaptive and switched beam antennas offer 10.63dB and 7.50dB respectively better user SINR compared to traditional wide antenna used in 3-sector site at 1000m intersite distance.

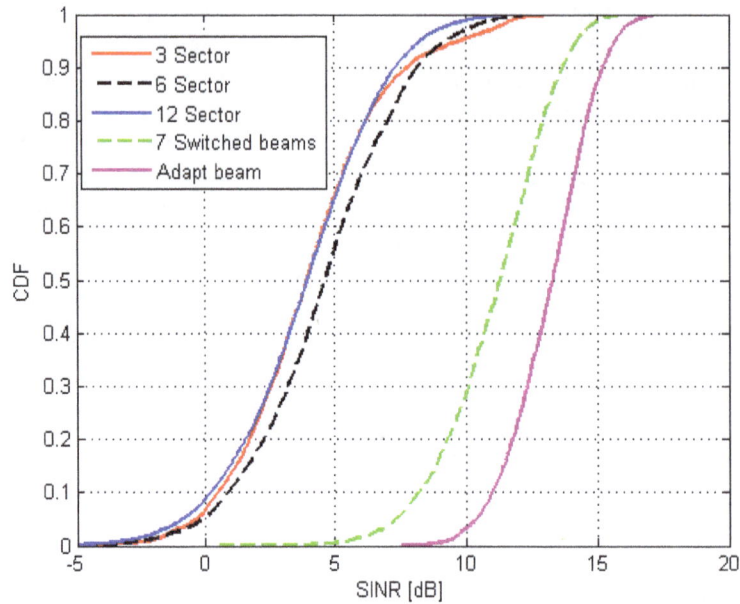

Fig.7. CDF plot of cell SINR with five users per TTI at 1000m ISD

Fig.7 shows the cumulative distribution function of SINR value averaged over the whole cell with 5 users per TTI at 1000m ISD for different simulated cases. Averaged SINR value over the whole cell area in each iteration of Monte Carlo simulation was obtained by adding the linear SINR value of each user and then divide the sum by number of users served per TTI. It can be seen that 6-sector deployment helps in improving the cell SINR by a small margin of 0.51dB only compared to 3-sector deployment, but a significant improvement of 7.02dB and 9.11dB is witnessed in case of switched beam and adaptive beam case respectively. Smart antennas not only improve the user experience rather they improve the overall cell SINR as well. It is also evident that the multiple switched beam antenna offers improvement in SINR but the difference is smaller compared to adaptive antenna. Statistical analysis of cell SINR is given in Table III.

Table III. Statistical Analysis of SINR over whole Cell

	10 percentile cell SINR (dB)	50 percentile cell SINR (dB)	90 percentile cell SINR (dB)	Mean cell SINR (dB)	STD cell SINR (dB)	Relative SINR gain (dB
3-Sector	0.49	3.91	7.79	4.07	2.94	**0**
6-Sector	0.91	4.63	8.14	4.58	2.85	**0.51**
12-Sector	0.21	3.94	7.31	3.84	2.80	**-0.23**
7 Switched beam	8.19	11.28	13.68	11.09	2.14	**7.02**
Adaptive beam	10.96	13.29	15.21	13.18	1.63	**9.11**

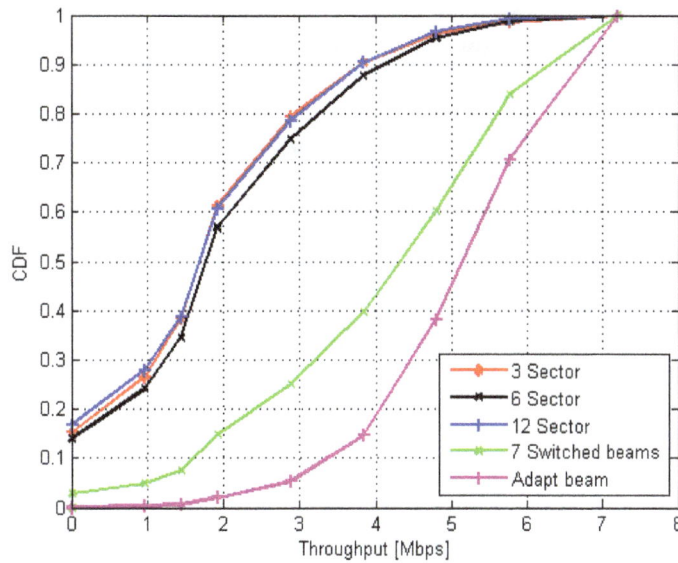

Fig.8. CDF plot of user throughput with 5 users per TTI at 1000m ISD

Fig.8. shows the CDF of the user throughput of DC-HSDPA network with 5 users per TTI at 1000m ISD for different antenna solutions. Eight marks on CDF plots represent eight different MCS. As equal codes and equal power was distributed among the users, therefore high throughput samples show that high modulation and coding scheme was used by the user. High MCS are less robust against interference and thus have high requirement of SINR. It is interesting to note that around 4.5% of the users were able to adapt 64QAM in 3-, 6-, and 12-sector case, whereas this number was raised to 39.98% and 61.8% by switched and adaptive beam antennas respectively. As seen from the results, more than 85% of the samples with adaptive beam were obtained with three highest MCS. Samples of zero throughputs in CDF plots represent the users with no data transfer due to very low SINR. It was also noted that single wide beam antenna keeps the probability of no data transfer at almost 15%. Whereas, switched beam antenna and adaptive beam antenna show remarkable improvement in probability of no data transfer and kept it at negligible level of 2.88% and 0.16% respectively. These results clearly indicate the impact of advanced antenna techniques in improving the user experience, when other cells are heavily loaded and are severely interfering the serving cell.

Fig.9 shows the CDF of cell throughput achieved by using DC-HSDPA with equal power and equal codes allocation for different network tessellation and antenna techniques. Cell throughput in each TTI was computed by summing the individual throughput of the served users. Like in previous results, case adaptive beam lead the comparison and shows extra ordinary performance compared to other network tessellations and antenna types in terms offering higher cell throughput. Almost identical cell throughput is achieved in 3-sector and 12-sector case, but 6-sector offers slightly better capacity. High SINR values showed in Fig.7 is translated into high throughput values in Fig.9. Adaptive beam antenna exhibits better performance and offers 27.99Mbps of average cell throughput compared to 22.81Mbps by switched beam case. 10 percentile cell throughput shows that 90% of the cell throughput samples with adaptive beam were above 24Mbps, and with switched beam 90% of the samples were above 17.28Mbps. Relative throughput gain is the relative gain in percentage value compared to 3-sector case. In [9], it was expected to get 100-200% improvement in cell capacity by smart adaptive antennas, and the results shown in Fig.9 are in line with the expectation.   Adaptive beam shows a significant relative gain of 156.7%, however switched beam have relative gain of 109.27%. Statistical analysis of cell throughput is shown in Table IV.

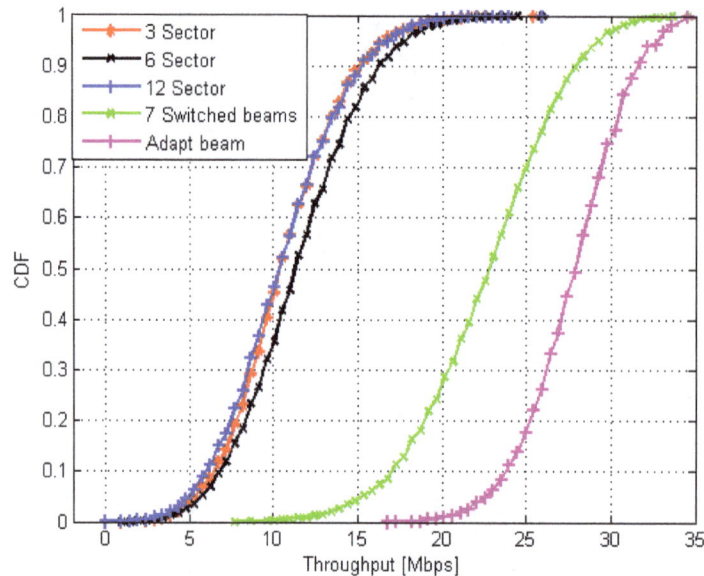

Fig.9. CDF plot of cell throughput with five users per TTI at 1000m ISD

Table IV. Statistical Analysis of Cell Throughput

	10 percentile cell throughput (Mbps)	50 percentile cell throughput (Mbps)	90 percentile cell throughput (Mbps)	Mean cell throughput (Mbps)	STD cell throughput (Mbps)	Relative throughput gain (%)
3-Sector	6.72	10.56	15.36	10.90	3.40	**0**
6-Sector	7.20	11.52	16.32	11.72	3.67	**7.52**
12-Sector	6.24	10.56	15.36	10.77	3.59	**-1.20**
7 Switched beams	17.28	23.04	28.32	22.81	4.23	**109.27**
Adaptive beam	24.0	28.32	31.68	27.99	3.03	**156.70**

Fig.10 shows the mean cell throughput of the DC-HSDPA cell with five users per TTI against the intersite distance for different cases. The trend of the sectored antenna cases and switched beam antenna case show that average cell throughput increases by increasing the intersite distance. Small intersite distance corresponds to small cells; hence, the high interference coming from the neighbor cells limit the cell throughput. The variations in the cell throughput for all cases except the adaptive antenna case were caused by the fact that larger the intersite distance, smaller will be the impact of interfering cells and hence larger will be the achieved average cell throughput. However, for adaptive antenna case cell throughput is inversely proportional to the intersite distance. The results show that a deployment of smart antennas significantly enhances the average cell throughput irrespective of the ISD. The highest cell throughput was achieved with adaptive beam antenna at small ISD of 250m. However, the worst capacity is offered by 12-sector antenna at 3000m ISD. It means higher order of sectorization not necessarily offers better cell throughput.

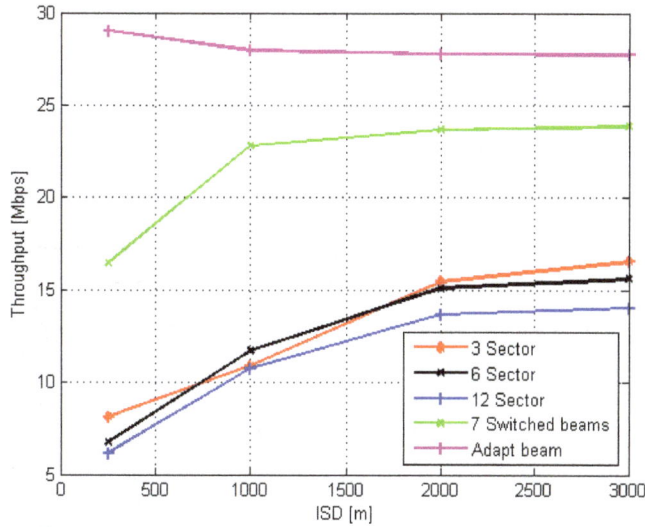

Fig.10. Mean cell throughput with five users per TTI against ISD

Fig.11 shows the achieved mean site throughput for DC-HSDPA system against the intersite distance for different cases. As seen in Fig.11, applying higher order sectorization and deploying advance antenna techniques provides significant throughput gain over traditional 3-sector site topology. Relative site throughput gain for 6-sector and 12-sector topology is higher at large intersite distances than small ISD. With respect to the reference case of 3-sector site at 1000m ISD, when intersite distance is reduced to 250m (small cell) for 3-sector site, mean site throughput is reduced by 25.41%. However, a relative throughput gain of approximately 23.67% and 125.69% is achieved by 6-sector and 12-sector sites respectively at 250m ISD, which is comparatively small compared to 164.04% and 401.65% by 6 and 12-sector sites respectively at 2000m ISD. Adaptive antenna beam outperformed at 250m ISD and was found more effective at small ISD. More detailed analysis of site throughput and the relative gain is presented in Table V. Relative gains shown in Table V are with respect to reference case 3-sector site at 1000m ISD. Negative value of gains means inferior performance.

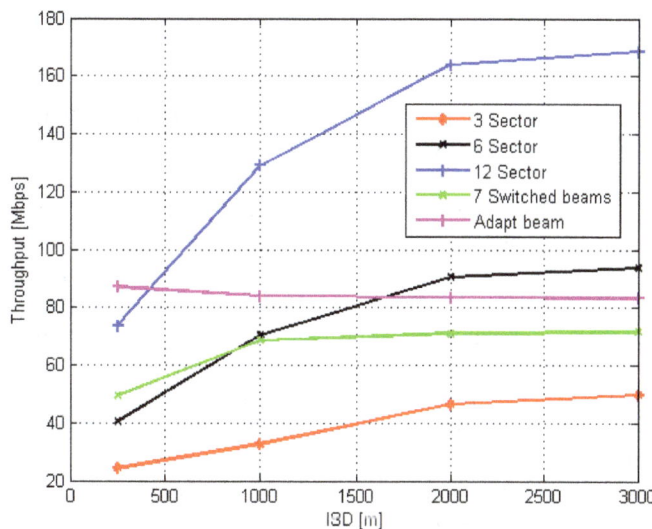

Fig.11. Mean site throughput with five users per TTI against ISD

Table V presents the average downlink throughput and relative sector (cell) gain with respect to 3-sector at 1000m ISD (reference case).

Table V. Statistical Analysis of Cell Throughput

	Mean cell throughput (Mbps)	Relative cell throughput gain (%)	Mean site throughput (Mbps)	Relative site throughput gain (%)
ISD = 250 meter				
3-Sector	8.13	**-25.41**	24.39	**-25.41**
7 Switched beams	16.48	**51.20**	49.44	**51.20**
Adaptive beam	29.01	**166.15**	87.03	**166.15**
6-Sector	6.74	**-38.17**	40.44	**23.67**
12-Sector	6.15	**-43.57**	73.80	**125.69**
ISD = 1000 meter as reference				
3-Sector	10.90	**0**	32.70	**0**
7 Switched beams	22.81	**109.27**	68.43	**109.27**
Adaptive beam	27.98	**156.70**	83.94	**156.70**
6-Sector	11.72	**7.52**	70.38	**115.23**
12-Sector	10.77	**-1.20**	129.24	**295.23**
ISD = 2000 meter				
3-Sector	15.49	**42.11**	46.47	**42.11**
7 Switched beams	23.71	**117.53**	71.13	**117.53**
Adaptive beam	27.80	**155.05**	83.40	**155.05**
6-Sector	15.10	**38.53**	90.60	**177.06**
12-Sector	13.67	**25.41**	164.04	**401.65**
ISD = 3000 meter				
3-Sector	16.58	**52.11**	49.74	**52.11**
7 Switched beams	23.89	**119.17**	71.67	**119.17**
Adaptive beam	27.75	**154.58**	83.25	**154.58**
6-Sector	15.63	**43.39**	93.78	**186.79**
12-Sector	14.06	**28.99**	168.72	**415.97**

6. CONCLUSION

In this article, we investigated advance antenna techniques along with different network tessellations including cloverleaf topology for 3-sector sites, snow flake topology for 6-sector sites and proposed a novel flower topology for 12-sector sites in DC-HSDPA network. Impact of intersite distance on the performance of higher order sectorization and on the performance of adaptive and switched beam antenna was also taken into account. A comprehensive set of simulation results were demonstrated together with a performance analysis. Post simulation analysis confirms that the capacity gain achieved by higher order sectorization and switched beam antenna increases by increasing the ISD. However, adaptive beam antenna also significantly improves the cell SINR and cell throughput, but adaptive antenna is more effective in small cells compared to large ISD. The simulation results revealed that the average cell SINR does not deteriorate much by having higher order sectorization, however 6-sector site provides around 0.5dB better cell SINR compared to 3-sector site. At 1000m ISD, the cell SINR is improved by approximately 7.02dB and 9.11dB when switched beam and adaptive beam antennas were used respectively compared to traditional 3-sector site with 65^0 beamwidth antenna. Significant improvement was also witnessed in terms of average cell throughput, it was found that the average cell throughput increased by 109.3% with multiple switched beam antenna, and 156.7%

with adaptive beam antenna compared to 3-sector site at 1000m ISD. Adaptive beam antenna outperformed and offered high SINR, high throughput with low probability of no data transfer. Multiple switched beam antenna showed better performance compared to single beam antenna but inferior to adaptive beam. Three-sector and higher order sectorization offer almost 15% of probability of no data transfer for the user at 1000m ISD. Switched beam antenna helps in improving the probability of no data transfer and kept it at almost 2.88%, but adaptive antenna significantly improved probability of no data transfer and brought it down to 0.16% at 1000m ISD. Simulaion results revealed that the user experience and the macro cell capacity can be significantly improved by deploying smart antennas. Higher order sectorization does not improve much the cell (sector) capacity, but definitely offers higher site capacity. Especially at large ISD, high order sectorization is more effective and significantly increases the site capacity. To avoid the deployment of small cells, usage of adaptive and switched beam antennas or higher order sectorization can be considered as an alternate choice.

The results were obtained by using semi-statistic simulations with Okumura-Hata propagation model, and thus may cause offset type of error in all results. However, the obtained results are still comparable with each other to show capacity improvements. For future work, it would be interesting to see the performance of fixed switched beam antenna with narrower and even more number of beams in a cell, as in this paper seven beams of 8^0 were considered in each cell.

ACKNOWLEDGEMENTS

Authors would like to thank Tampere University of Technology, European Communications Engineering (ECE) Ltd. and Tampere Doctoral Program of Information Science and Engineering (TISE) for supporting the research work of this paper.

REFERENCES

[1]   W.C.Y. Lee, "Mobile communications design fundamentals", John Wiley & Sons, Inc., 1993.
[2]   L. C. Wang, K. Chawla, L. J. Greenstein, "Performance Studies of Narrow-Beam Trisector Cellular Systems," International Journal of Wireless Information Networks, Vol. 5, No. 2, 1998, pp. 89-102.
[3]   I.V. Stevanović, A. Skrivervik, J. R. Mosig, "Smart antenna systems for mobile communications". Laboratoire d'Electromagnetisme et d'Acoustique Ecole Polytechnique Federale de Lausanne, January 2003.
[4]   A. Wacker, J. Laiho-Steffens, K.Sipilä, K. Heiska, "The impact of the base station sectorisation on WCDMA radio network performance," in Proc. 50th IEEE Vehicular Technology Conference, 1999, pp. 2611-2615.
[5]   B. C. V. Johansson, S. Stefansson, "Optimizing Antenna Parameters for Sectorized W-CDMA Networks," in Proc. IEEE Vehicular Technology Conference, 2000, pp. 1524-1531.
[6]   J. Niemelä, T. Isotalo, J. Lempiäinen, "Optimum Antenna Downtilt Angles for Macrocellular WCDMA Network," EURASIP Journal on Wireless Communications and Networking, vol. 5, 2005.
[7]   J. Itkonen, B. Tuszon, J. Lempiäinen, "Assessment of network layouts for CDMA radio access", EURASIP Journal on Wireless Communications and Networking, Volume 2008
[8]   J. Lempiäinen, M. Manninen, "Radio interface system planning for GSM/GPRS/UMTS", Kluwer Academic Publishers. 2001.
[9]   C.S. Rani., P.V. Subbaiah., K.C. Reddy, "LMS and RLS Algorithms for smart antennas in a CDMA mobile communication environment". International Journal of the Computer the Internet and Management (IJCIM), Vol.16 No.2, May-August 2008.
[10] E. Dahlman, S. Parkvall, J. Skold, "3G Evolution: HSPA and LTE for mobile broadband", Academic press, First edition, 2007.
[11] K. Johansson, J. Bergman, D. Gerstenberger, M. Blomgren, A. Wallen, "Multi-carrier HSPA evolution", Vehicular Technology Conference, 2009. VTC Spring 2009. IEEE 69th , pp.1-5, 26-29 April 2009.

[12] D. Soldani, M. Li, and R. Cuny, "QoS and QoE management in UMTS cellular systems". John Wiley & Sons Ltd, 2006.

[13] A. Chheda, F. Bassirat, " Enhanced cellular network layout for CDMA networks having six-sectored cells ", U.S.Patent 5960349, 28th September 1999.

[14] J. D. Kraus, R. J. Marhefka, "Antennas for all applications". 3rd ed. 2001, McGraw-Hill.

[15] R. Kawitkar, "Issues in deploying smart antennas in mobile radio networks," Proceedings of World Academy of Science, Engineering and Technology Vol.31, July 2008, pp. 361-366, ISSN 1307-6884.

[16] A. F. Molisch, "Wireless communications". 2nd edition. UK 2011, John Wiley & Sons Ltd.

[17] D. Cabrera, J. Rodriguez, "Switched beam smart antenna BER performance analysis for 3G CDMA cellular communication", Computer Research Conference CRC2004, Puerto Rico, April 2004.

[18] A.U. Bhobe, P.L. Perini, "An overview of smart antenna technology for wireless communication", IEEE Aerospace Conference, Vol.2, pp. 875-883, 2001.

[19] C.S. Rani., P.V. Subbaiah., K.C. Reddy, "Smart antenna algorithms for WCDMA mobile communication systems", International Journal of Computer Science and Network Security (IJCSNS), Vol.8 No. 7, July 2008.

[20] S. Hossain, M.T. Islam, S. Serikawal, "Adaptive beamforming algorithms for smart antenna systems", International conference on control, automation and systems, ICCAS 2008. p. 412–416.

Permissions

All chapters in this book were first published in IJWMN, by AIRCC Publishing Corporation; hereby published with permission under the Creative Commons Attribution License or equivalent. Every chapter published in this book has been scrutinized by our experts. Their significance has been extensively debated. The topics covered herein carry significant findings which will fuel the growth of the discipline. They may even be implemented as practical applications or may be referred to as a beginning point for another development.

The contributors of this book come from diverse backgrounds, making this book a truly international effort. This book will bring forth new frontiers with its revolutionizing research information and detailed analysis of the nascent developments around the world.

We would like to thank all the contributing authors for lending their expertise to make the book truly unique. They have played a crucial role in the development of this book. Without their invaluable contributions this book wouldn't have been possible. They have made vital efforts to compile up to date information on the varied aspects of this subject to make this book a valuable addition to the collection of many professionals and students.

This book was conceptualized with the vision of imparting up-to-date information and advanced data in this field. To ensure the same, a matchless editorial board was set up. Every individual on the board went through rigorous rounds of assessment to prove their worth. After which they invested a large part of their time researching and compiling the most relevant data for our readers.

The editorial board has been involved in producing this book since its inception. They have spent rigorous hours researching and exploring the diverse topics which have resulted in the successful publishing of this book. They have passed on their knowledge of decades through this book. To expedite this challenging task, the publisher supported the team at every step. A small team of assistant editors was also appointed to further simplify the editing procedure and attain best results for the readers.

Apart from the editorial board, the designing team has also invested a significant amount of their time in understanding the subject and creating the most relevant covers. They scrutinized every image to scout for the most suitable representation of the subject and create an appropriate cover for the book.

The publishing team has been an ardent support to the editorial, designing and production team. Their endless efforts to recruit the best for this project, has resulted in the accomplishment of this book. They are a veteran in the field of academics and their pool of knowledge is as vast as their experience in printing. Their expertise and guidance has proved useful at every step. Their uncompromising quality standards have made this book an exceptional effort. Their encouragement from time to time has been an inspiration for everyone.

The publisher and the editorial board hope that this book will prove to be a valuable piece of knowledge for researchers, students, practitioners and scholars across the globe.

List of Contributors

Mrs. M.B. VEENA
Research scholar, ECE department, SJCE, Mysore. Karnataka, India

Dr. M.N. SHANMUKHA SWAMY
Professor ECE department, SJCE, Mysore. Karnataka, India

P. Priya Naidu
Associate System Engineer, IBM Pvt. Limited, India

Meenu Chawla
Associate Professor, Department of computer Science MANIT, Bhopal, India

Justin James
ARO Center for Battlefield Communications, Department of Electrical and Computer Engineering, Prairie View A&M University, Texas 77446

Annamalai Annamalai
ARO Center for Battlefield Communications, Department of Electrical and Computer Engineering, Prairie View A&M University, Texas 77446

Olusegun Odejide
ARO Center for Battlefield Communications, Department of Electrical and Computer Engineering, Prairie View A&M University, Texas 77446

Dhadesugoor Vaman
ARO Center for Battlefield Communications, Department of Electrical and Computer Engineering, Prairie View A&M University, Texas 77446

K. Thilagam
Research Scholar, Department of ECE, Pondicherry Engg. College, Puducherry, India

K. Jayanthi
Associate Professor, Department of ECE, Pondicherry Engg. College, Puducherry,India

Manjusha Pandey
Indian Institute of Information Technology Allahabad, India

Shekhar Verma
Indian Institute of Information Technology Allahabad, India

Mehdi Vasef
University of Duisburg-Essen, Germany

R. Bhakthavathsalam
Supercomputer Education and Research Centre Indian Institute of Science, Bangalore, India

Khurram J. Mohammed
Ghousia College of Engineering, Ramanagaram, India

Shahid Md. Asif Iqbal
Department of Computer Science & Engineering, Premier University, Chittagong, Bangladesh

Md. Humayun Kabir
Department of Computer Science and Engineering, Bangladesh University of Engineering and Technology, Dhaka, Bangladesh

Bassam A. Alqaralleh
Al-Hussein Bin Talal University, Jordan

Khaled Almi'ani
Al-Hussein Bin Talal University, Jordan

John Tengviel
Department of Computer Science, Sunyani Polytechnic, Sunyani, Ghana

K. Diawuo
Department of Computer Engineering, KNUST, Kumasi, Ghana

Debanjan Das Deb
Department of Computer Science & Engineering, B. P. Poddar Institute of Management & Technology, India

Sagar Bose
PervCom Consulting Pvt. Ltd., India

Somprakash Bandyopadhyay
Indian Institute of Management Calcutta, India

Anuj Vadhera and Lavish Kansal
Lovely Professional University, Phagwara, Punjab, India

Muhammad Usman Sheikh
Department of Communications Engineering, Tampere University of Technology

Jukka Lempiainen
Department of Communications Engineering, Tampere University of Technology

Hans Ahnlund
European Communications Engineering Ltd, Tekniikantie 12, Espoo Finland